Flow-Based Optimization of Products or Devices

Flow-Based Optimization of Products or Devices

Special Issue Editor
Nils Tångefjord Basse

MDPI • Basel • Beijing • Wuhan • Barcelona • Belgrade • Manchester • Tokyo • Cluj • Tianjin

Special Issue Editor
Nils Tångefjord Basse
Independent Scholar
Sweden

Editorial Office
MDPI
St. Alban-Anlage 66
4052 Basel, Switzerland

This is a reprint of articles from the Special Issue published online in the open access journal *Fluids* (ISSN 2311-5521) (available at: https://www.mdpi.com/journal/fluids/special_issues/flow_optimization).

For citation purposes, cite each article independently as indicated on the article page online and as indicated below:

LastName, A.A.; LastName, B.B.; LastName, C.C. Article Title. *Journal Name* **Year**, *Article Number*, Page Range.

ISBN 978-3-03936-441-1 (Hbk)
ISBN 978-3-03936-442-8 (PDF)

© 2020 by the authors. Articles in this book are Open Access and distributed under the Creative Commons Attribution (CC BY) license, which allows users to download, copy and build upon published articles, as long as the author and publisher are properly credited, which ensures maximum dissemination and a wider impact of our publications.

The book as a whole is distributed by MDPI under the terms and conditions of the Creative Commons license CC BY-NC-ND.

Contents

About the Special Issue Editor vii

Nils T. Basse
Flow-Based Optimization of Products or Devices
Reprinted from: *Fluids* **2020**, *5*, 56, doi:10.3390/fluids5020056 1

Joe Alexandersen and Casper Schousboe Andreasen
A Review of Topology Optimisation for Fluid-Based Problems
Reprinted from: *Fluids* **2020**, *5*, 29, doi:10.3390/fluids5010029 7

Palanisamy Mohan Kumar, Mohan Ram Surya, Krishnamoorthi Sivalingam, Teik-Cheng Lim, Seeram Ramakrishna and He Wei
Computational Optimization of Adaptive Hybrid Darrieus Turbine: Part 1
Reprinted from: *Fluids* **2019**, *4*, 90, doi:10.3390/fluids4020090 39

Brice Rogie, Wiebke Brix Markussen, Jens Honore Walther and Martin Ryhl Kærn
Numerical Investigation of Air-Side Heat Transfer and Pressure Drop Characteristics of a New Triangular Finned Microchannel Evaporator with Water Drainage Slits
Reprinted from: *Fluids* **2019**, *4*, 205, doi:10.3390/fluids4040205 59

Pavlos Alexias and Kyriakos C. Giannakoglou
Shape Optimization of a Two-Fluid Mixing Device Using Continuous Adjoint
Reprinted from: *Fluids* **2020**, *5*, 11, doi:10.3390/fluids5010011 81

Micaela Olivetti, Federico Giulio Monterosso, Gianluca Marinaro, Emma Frosina and Pietro Mazzei
Valve Geometry and Flow Optimization through an Automated DOE Approach
Reprinted from: *Fluids* **2020**, *5*, 17, doi:10.3390/fluids5010017 97

Michael Parker and Douglas Bohl
Experimental Investigation of Finite Aspect Ratio Cylindrical Bodies for Accelerated Wind Applications
Reprinted from: *Fluids* **2020**, *5*, 25, doi:10.3390/fluids5010025 117

Shenan Grossberg, Daniel S. Jarman and Gavin R. Tabor
Derivation of the Adjoint Drift Flux Equations for Multiphase Flow
Reprinted from: *Fluids* **2020**, *5*, 31, doi:10.3390/fluids5010031 135

Joel Guerrero, Luca Mantelli and Sahrish B. Naqvi
Cloud-Based CAD Parametrization for Design Space Exploration and Design Optimization in Numerical Simulations
Reprinted from: *Fluids* **2020**, *5*, 36, doi:10.3390/fluids5010036 157

About the Special Issue Editor

Nils Tångefjord Basse received his Ph.D. in 2002 from the Niels Bohr Institute, University of Copenhagen, on optical turbulence measurements in fusion plasmas. He was a Postdoctoral Associate at MIT, where he continued research on plasma turbulence. In 2006, he transitioned to corporate research on circuit breakers and worked as a Scientist/Principal Scientist at ABB in Switzerland. He went on to work as a Senior Research Engineer at Siemens in Denmark in the field of flowmeters, followed by industrial research on valves at Danfoss. Currently, he is a Senior CAE Engineer at Volvo Cars in Sweden, focusing on thermofluid simulations of electrical machines.

Editorial

Flow-Based Optimization of Products or Devices

Nils T. Basse

Elsas väg 23, 423 38 Torslanda, Sweden; nils.basse@npb.dk

Received: 14 April 2020; Accepted: 17 April 2020; Published: 22 April 2020

Keywords: flow-based optimization; internal and/or external flow; modelling and simulation; computational fluid dynamics; measurements and theory

1. Introduction

Flow-based optimization of products and devices is an immature field compared to corresponding topology optimization based on solid mechanics. However, it is an essential part of component development with both internal and/or external flow.

Flow-based optimization can be achieved by e.g., coupling of computational fluid dynamics (CFD) and optimization software; both open-source and commercial options exist. The motivation for flow-based optimization can be to improve performance, reduce size/cost, extract additional information or a combination of these objectives. The outcome of the optimization process may be geometries which are more suitable for additive manufacturing (AM) instead of traditional subtractive manufacturing.

This MDPI Fluids Special Issue (SI) is a two-fold effort to:

- Provide state-of-the-art examples of flow-based optimization; Table 1 contains an overview of the topics treated in this SI. Also included are the various Quantities of Interest (QoI).
- Present "A Review of Topology Optimisation for Fluid-Based Problems" by Alexandersen and Andreasen [1].

Table 1. Overview of Special Issue research contributions: Applications and Quantities of Interest.

Paper	Application	Quantities of Interest
Kumar et al. [2]	Wind turbine	Power and torque coefficients
Rogié et al. [3]	Microchannel evaporator	Heat transfer and pressure drop
Alexias et al. [4]	Longer static mixing device	Mixture uniformity and pressure drop
Alexias et al. [4]	Shorter static mixing device	Mixture uniformity and pressure drop
Olivetti et al. [5]	Valve	Mass flow rate
Parker et al. [6]	Accelerated wind bodies	Pressure coefficient and velocity
Grossberg et al. [7]	Dispersed multiphase flow	Mass flow rate
Guerrero et al. [8]	Cylinder	Surface area
Guerrero et al. [8]	Static mixer	Velocity distribution
Guerrero et al. [8]	Ahmed bodies	Normalized drag coefficient

2. Research

The research papers are briefly introduced in chronological order; methods and tools applied are summarized in Table 2. Note that all CFD simulations are steady-state and that all simulation-based methods include CAD-based operations to some extent.

Table 2. Overview of the Special Issue research contributions: Optimization methods and tools. Abbreviations: Computational Fluid Dynamics (CFD), Design of Experiments (DoE), Design Space Exploration (DES) and Design Optimization (DO).

Paper	Methods	Tools
Kumar et al. [2]	Parametric optimization	2D CFD: Turbulent flow
Rogié et al. [3]	Parametric optimization	3D CFD: Turbulent flow
Alexias et al. [4]	Continuous adjoint	3D CFD: Laminar flow
Olivetti et al. [5]	Automated DoE	Optimization tool and 3D CFD: Turbulent flow
Parker et al. [6]	Smooth and corrugated cylinder	Measurements of pressure and velocity
Grossberg et al. [7]	Continuous adjoint	Derivation of the adjoint drift flux equations
Guerrero et al. [8]	Cloud-based DSE and DO	Optimization tool and 3D CFD: Turbulent flow

The paper by Kumar et al. [2] is on the topic of small-scale decentralized wind power generation; the authors propose an adaptive hybrid Darrieus turbine (AHDT) to overcome issues experienced by Savonius and Darrieus wind turbines. The AHDT has a Savonius rotor nested inside a Darrieus rotor, where the Savonius rotor can change shape. Optimization consists of changing the diameter of the Savonius rotor while keeping the Darrieus rotor diameter fixed. 2D CFD simulations using the $k-\omega$ shear-stress transport (SST) turbulence model are carried out to study the hybrid turbine performance. The torque coefficient is optimized, which is defined as the ratio of generated aerodynamic torque to the available torque in the wind. The corresponding power coefficient for different tip speed ratios is also characterized. Flow interaction between the Savonius rotor in closed configuration and the Darrieus rotor blades takes place due to the formation of Kármán vortices.

Rogié et al. [3] compare new microchannel evaporator designs to a baseline finned-tube evaporator; the new designs have drainage slits for improved moisture removal with triangular shaped plain fins. Optimization is carried out by varying the geometry (transverse tube pitch and triangular fin pitch) and the inlet velocity while keeping a constant wall temperature of tube and fin. 3D $k-\omega$ SST CFD simulations were done to establish heat transfer coefficients and pressure drop, both as a function of tube rows. These results were in turn used to develop Colburn j-factor and Fanning f-factor correlations. It was found that the entrance region is very important for heat transfer and that the new designs transfer more heat per unit volume than the baseline.

The continuous adjoint method is applied by Alexias and Giannakoglou [4] to study two-fluid mixing devices. The authors consider laminar flow of two miscible fluids and change baffle shapes and angles to optimize (i) mixture uniformity at the exit and (ii) the total pressure loss occurring between the inlets and the outlet. These two objectives are used to construct a single target function. The primal (flow) and adjoint field equations are solved and thereafter the sensitivity derivatives are found. Two mixing devices are treated, one longer (with 7 baffles) and one shorter (with 4 baffles). Both have two inlets and one outlet. Three optimization scenarios are tested using combinations of node-based parametrization (NBP) and positional angle parametrization (PAP). Results are presented and it is demonstrated that the shorter mixing device has a lower pressure drop but also worse mixing quality than the longer mixing device.

Olivetti et al. [5] optimize a four-way hydropiloted valve by combining an optimization tool (with integrated parametric geometry) and CFD simulations. The 3D CFD simulations uses the standard $k-\varepsilon$ turbulence model. The shape of two ports of the valve are optimized to maximise mass flow rate for a fixed static pressure difference between the two ports. A Design of Experiments (DoE) sequence is generated with a Sobol algorithm which determined that 8 design variables resulting in 90 variants should be simulated. The Sobol sequence resulted in a significant increase of the mass flow rate. A second optimization step was done on the best Sobol sequence design using a 2-level tangent search (Tsearch) method which led to a further improvement. Experiments confirmed the findings obtained using the CFD-based optimization.

Cylindrical bodies for "accelerated wind" applications are experimentally characterized by Parker and Bohl [6]. Here, one aims to enhance power extraction from wind by adding a structure near the rotor to increase the flow velocity, i.e., to increase the kinetic energy of the wind before it reaches the wind turbine blades. Two short aspect ratio cylindrical bodies are tested, a corrugated and a smooth cylinder. The cylindrical bodies are tested in a wind tunnel using varying Reynolds number (Re); pressure taps are placed in the bodies and the velocity is measured with hot-wire probes. End effects are found to be important. Both bodies demonstrated increased flow speed, but gauged by the pressure coefficient and velocity, the smooth cylinder exhibited better performance than the corrugated cylinder.

The continuous adjoint method is applied to dispersed multiphase systems by Grossberg et al. [7]. A drift-flux model is studied, where the two separate phases are considered as a single mixture phase. This is a simplification compared to the two-fluid formulation. The transport of the dispersed phase is modelled using a drift equation; this equation, along with mixture-momentum and mixture-continuity equations, forms the drift flux (primal) equations. The adjoint drift flux equations with a Darcy porosity term are derived under the frozen turbulence (or constant mixture turbulent viscosity) assumption. The corresponding boundary conditions for the adjoint variables are also calculated. Application examples are documented for wall-bounded flows, where (i) adjoint boundary conditions, (ii) the objective function and (iii) the settling (drift) velocity are derived. The objective function is the mass flow rate of the dispersed phase at the outlet.

Guerrero et al. [8] present an engineering design framework with a cloud-based parametrical CAD application which can be used on any platform without the need for a local installation. The optimization loop is fault-tolerant and scalable in the sense that both concurrent and parallel simulations can be deployed. Two methods are used for optimization: Design Space Exploration (DSE) and Design Optimization (DO). DO converges to an optimal design, either using a (i) gradient-based or (ii) derivative-free method. In contrast, DSE is used to explore the design space in a methodical fashion without converging to an optimum. Results from DSE provide more information to the engineer than DO and can also be used for e.g., surrogate-based optimization studies which are orders of magnitude faster than working at the high fidelity level. A useful approach can be to carry out a DSE as a first step, followed by a DO. Three numerical experiments are documented in the paper: The first example minimizes the total surface area of a cylinder with a given volume and serves to introduce the optimization framework. The second example on a static mixer uses 3D CFD simulations with the $k - \varepsilon$ turbulence model and compares velocity profile images using the Structural Similarity Index (SSIM) method. The third example is on changing the inter-vehicle spacing between two Ahmed bodies to calculate the resulting normalized drag coefficient. 3D CFD using the $k - \omega$ SST turbulence model is used and the simulations are compared to measurements.

3. Review

Alexandersen and Andreasen [1] have written the first complete review on topology optimization for fluid-based problems. This research field was started in 2003; at that point in time, topology optimization of solid mechanics had already been an active research area for 15 years. 186 papers are covered by the literature review according to the selection criterion that at least one governing equation for fluid flow must be solved; the topics are summarized in Table 3.

The quantitative analysis of the literature discusses the total number of publications per year and how these are distributed in terms of:

- Design representations, e.g., density-based and level set methods
- Discretization methods, e.g., the finite element method and the lattice Boltzmann method
- Problem types, e.g., pure fluid and conjugate heat transfer
- Flow types, e.g., steady-state and transient laminar flow
- Dimensionality, i.e., 2D or 3D

Recommendations are given, ranging from methods used, to which types of physical problems the community should focus on in the future. Topics covered by the recommendations include:

- Optimization methods
- Density-based approaches
- Level set-based approaches
- Steady-state laminar incompressible flow
- Benchmarking
- Time-dependent problems
- Turbulent flow
- Compressible flow
- Fluid-structure interaction
- 3D problems
- Simplified models or approximations
- Numerical verification
- Experimental validation

Finally, to quote from the Conclusions of the paper, "The community is encouraged to focus on moving the field to more complicated applications, such as transient, turbulent and compressible flows."

Table 3. Overview of flow topics treated in the review paper.

Main Topic	Subtopic (If Applicable)
Fluid flow	Steady laminar flow
	Unsteady flow
	Turbulent flow
	Non-Newtonian fluids
Species transport	
Conjugate heat transfer	Forced convection
	Natural convection
Fluid-structure interaction	
Microstructure and porous media	Material microstructures
	Porous media

4. Conclusions

Examples of flow-based optimization research have been provided in this Special Issue along with a complete review of the research field.

There is a natural connection between flow-based optimization and AM, since geometrical shapes resulting from optimization may be challenging to realize using traditional manufacturing methods. Note that Parker and Bohl [6] used AM to manufacture the corrugated cylinder. We recommend researchers in the field to use AM more extensively in the future to test geometries from simulation studies.

Another area where more synergy can be explored is to combine Design Space Exploration and Machine Learning [9,10] as is also mentioned by Guerrero et al. [8].

A range of physical flow phenomena which are suitable for topology optimization exists, see e.g., the list in the Special Issue Information Section [11]. We look forward to following the research field in the future; surely, this is only the beginning!

Acknowledgments: We would like to thank all the authors who contributed to this Special Issue. We are also grateful to all the anonymous reviewers for their help; without the help of qualified reviewers, it would not have been possible to organize this Special Issue. A personal note of appreciation and gratitude to Sonia Guan, the Managing Editor of Fluids, and the editorial staff at the Fluids Office; without their help and assistance, Fluids could not publish high quality papers in a short period of time.

Conflicts of Interest: The author declares no conflict of interest.

References

1. Alexandersen, J.; Andreasen, C.S. A Review of Topology Optimisation for Fluid-Based Problems. *Fluids* **2020**, *5*, 29. [CrossRef]
2. Kumar, P.M.; Surya, M.R.; Sivalingam, K.; Lim, T.-C.; Ramakrishna, S.; Wei, H. Computational Optimization of Adaptive Hybrid Darrieus Turbine: Part 1. *Fluids* **2019**, *4*, 90. [CrossRef]
3. Rogié, B.; Markussen, W.B.; Walther, J.H.; Kærn, M.R. Numerical Investigation of Air-Side Heat Transfer and Pressure Drop Characteristics of a New Triangular Finned Microchannel Evaporator with Water Drainage Slits. *Fluids* **2019**, *4*, 205. [CrossRef]
4. Alexias, P.; Giannakoglou, K.C. Shape Optimization of a Two-Fluid Mixing Device Using Continuous Adjoint. *Fluids* **2020**, *5*, 11. [CrossRef]
5. Olivetti, M.; Monterosso, F.G.; Marinaro, G.; Frosina, E.; Mazzei, P. Valve Geometry and Flow Optimization through an Automated DOE Approach. *Fluids* **2020**, *5*, 17. [CrossRef]
6. Parker, M.; Bohl, D. Experimental Investigation of Finite Aspect Ratio Cylindrical Bodies for Accelerated Wind Applications. *Fluids* **2020**, *5*, 25. [CrossRef]
7. Grossberg, S.; Jarman, D.S.; Tabor, G.R. Derivation of the Adjoint Drift Flux Equations for Multiphase Flow. *Fluids* **2020**, *5*, 31. [CrossRef]
8. Guerrero, J.; Mantelli, L.; Naqvi, S.B. Cloud-Based CAD Parametrization for Design Space Exploration and Design Optimization in Numerical Simulations. *Fluids* **2020**, *5*, 36. [CrossRef]
9. MDPI Fluids Special Issue on "Numerical Fluid Flow Simulation Using Artificial Intelligence and Machine Learning". Available online: https://www.mdpi.com/journal/fluids/special_issues/Artificial_Intelligence_and_Machine_Learning (accessed on 21 April 2020).
10. Brunton, S.L.; Noack, B.R.; Koumoutsakos, P. Machine Learning for Fluid Mechanics. *Annu. Rev. Fluid Mech.* **2020**, *52*, 477–508. [CrossRef]
11. MDPI Fluids Special Issue on "Flow-Based Optimization of Products or Devices". Available online: https://www.mdpi.com/journal/fluids/special_issues/flow_optimization (accessed on 21 April 2020).

© 2020 by the author. Licensee MDPI, Basel, Switzerland. This article is an open access article distributed under the terms and conditions of the Creative Commons Attribution (CC BY) license (http://creativecommons.org/licenses/by/4.0/).

Review

A Review of Topology Optimisation for Fluid-Based Problems

Joe Alexandersen [1],* and Casper Schousboe Andreasen [2]

1 Department of Technology and Innovation, University of Southern Denmark, 5230 Odense, Denmark
2 Department of Mechanical Engineering, Technical University of Denmark, 2800 Kgs. Lyngby, Denmark; csan@mek.dtu.dk
* Correspondence: joal@iti.sdu.dk

Received: 7 February 2020; Accepted: 27 February 2020; Published: 4 March 2020

Abstract: This review paper provides an overview of the literature for topology optimisation of fluid-based problems, starting with the seminal works on the subject and ending with a snapshot of the state of the art of this rapidly developing field. "Fluid-based problems" are defined as problems where at least one governing equation for fluid flow is solved and the fluid–solid interface is optimised. In addition to fluid flow, any number of additional physics can be solved, such as species transport, heat transfer and mechanics. The review covers 186 papers from 2003 up to and including January 2020, which are sorted into five main groups: pure fluid flow; species transport; conjugate heat transfer; fluid–structure interaction; microstructure and porous media. Each paper is very briefly introduced in chronological order of publication. A quantititive analysis is presented with statistics covering the development of the field and presenting the distribution over subgroups. Recommendations for focus areas of future research are made based on the extensive literature review, the quantitative analysis, as well as the authors' personal experience and opinions. Since the vast majority of papers treat steady-state laminar pure fluid flow, with no recent major advancements, it is recommended that future research focuses on more complex problems, e.g., transient and turbulent flow.

Keywords: topology optimisation; review paper; fluid flow; multiphysics; species transport; conjugate heat transfer; fluid–structure interaction; porous media

---

1. Introduction

The topology optimisation method originates from the field of solid mechanics, where it emerged from sizing and shape optimisation by the end of the 1980s. The seminal paper on topology optimisation is often quoted as being the homogenisation method by Bendsøe and Kikuchi [1]. Topology optimisation is posed as a material distribution technique that answers the question "where should material be placed?" or alternatively "where should the holes be?". As a structural optimisation method, it distinguishes itself from the more classical disciplines of sizing and shape optimisation, by the fact that there does not need to be an initial structure defined a priori. Having stated that, we define topology optimisation slightly wider in this context, as we include optimisation approaches in which the topology is allowed to or can change during the optimisation process. The review papers by Sigmund and Maute [2] and Deaton and Grandhi [3] give a general overview of topology optimisation methods and applications. Today, topology optimisation for solid mechanics is a mature technology that is widely available in all major finite element analysis (FEA) packages and even in many computer aided design (CAD) packages. The technology is utilised at the component design level in the automotive and aerospace industries.

The ideas of the original methodology are extendable to all physics, where the governing equations can be described by a set of partial differential equations (PDEs). It has therefore in the post-2000 decades seen widespread application to a range of different physics, such as acoustics, photonics, electromagnetism, heat conduction, fluid flow, etc. [3].

When applied to fluid problems, the question should be rephrased from "where should the holes be?" to "where should the fluid flow?". The optimisation problem basically becomes a question of where to enforce relevant boundary conditions for the flow problem. This review paper is a survey of published papers containing topology optimisation of fluid flow problems and related fluid-based problems. It is the first to cover the entire history, from its very beginning to the current state of the art. There are two previous review papers dealing with two different subsets under the umbrella of fluid-based problems, namely microfluidics [4] and thermofluidics [5].

1.1. Definitions for Inclusion

In the following, the scope and limitations of the review and the applied definition for *fluid-based problems* are elaborated upon.

1.1.1. Governing Equations

The solved problems must include fluid flow, meaning that at least one governing equation for fluid flow must be solved, such as:

- Darcy, Forchheimer and Brinkman flow
- Stokes and Navier–Stokes flow
- Homogenised fluid equations
- Kinetic gas theory, Lattice Boltzmann and similar methods based on distributions
- Particle methods

Therefore, papers treating only hydrostatic fluid loading (Laplace equation for pressure) and acoustics (Helmholtz equation for sound pressure) are omitted. In addition to this, for fluid–structure interaction problems, where only the structural part is optimised (so-called "dry optimisation") has also been left out.

In addition to the governing equations for fluid flow, any number of additional physics can be solved. The additional physics can be uncoupled, loosely coupled (one-way) or fully coupled (two-way), as long as a fluid problem is included in the optimisation formulation in the form of the objective functional or constraints. Examples are:

- Species transport, e.g., microfluidic mixers,
- Reaction kinetics, e.g., ion transport in flow batteries,
- Temperature, e.g., heat exchangers,
- Structural mechanics, e.g., fluid–structure interaction.

1.1.2. Literature Search

In order to collect relevant literature for this review, a literature search was performed using Google Scholar based on the keyword combinations of "topology optimization" with the following:

- fluid flow
- conjugate heat transfer
- convection
- fluid structure interaction
- microstructure
- homogenization

In addition to the above, reverse tracking was used of citations of the seminal papers in the area, as well as relevant references in the papers from the search. Only journal publications have been included, except when important contributions have been made in available conference proceedings.

1.1.3. Optimisation Methodology

A broad and open definition of *topology optimisation* is used herein. The presented methodologies must be capable of handling topological changes in three dimensions. That is, the methods should in a three-dimensional version be capable of handling large design changes and topological changes by *creating*, *removing* and *merging* holes. This can be difficult for some representations, especially in two dimensions, where auxiliary information such as topological derivatives is necessary for the creation of holes/structure. Pure shape optimisation, where only small modifications of the fluid–solid or fluid–void interface is possible, is not included in this review.

However, there might not be much need for changes in topology for most two-dimensional fluid flow problems. Due to the nature of fluid flow and the obvious objective of minimising power dissipation (or pressure drop), there is a desire to minimise the number of flow channels, i.e., only a single flow path is needed and the interface shape is modified. The need for topology optimisation does arise when other objectives, such as flow uniformity or diodicity, are considered and when the fluid flow is coupled to additional physics, such as e.g., heat transport.

The design representation used for topology optimisation of fluid-based problems is in general similar to those applied within the area of solid mechanics [2,6]. Figure 1 shows the three general options for representing the design. The first representation is an explicit boundary representation based on a body-fitted mesh adopting to the nominal geometry shown in red. If the design is changed, boundary nodes must be moved and the mesh must be updated or the domain must be entirely re-meshed if large changes are applied. The second representation is that of the density-based methods, which also includes level set methods where a smooth Heaviside projection is applied together with interpolation of material properties (so-called Ersatz material methods). The flow is penalised in the solid (black) domain, typically by modelling it as a porous material with very low permeability. The third representation is that of surface-capturing level set methods, where surface-capturing discretisation methods, e.g., the extended finite element method (X-FEM), where the cut elements are integrated using a special scheme and the interface boundary conditions are imposed, e.g., using stabilised Lagrange multipliers or a stabilised Nitsche's method.

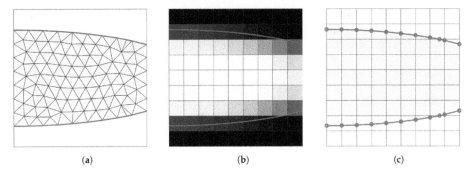

Figure 1. Fluid nozzle illustrating the basic differences among design representations in topology optimisation: (**a**) explicit boundary representation (body fitted mesh); (**b**) density/ersatz material based representation; (**c**) level set based X-FEM/cutFEM representation.

The explicit boundary methods represent the physics well, but moving nodes and adaption of the mesh are non-smooth operations and this might pose difficulties in advancing the design for the optimiser. Furthermore, regularisation of the interface is necessary and, in case of full re-meshing, it might be difficult to assure high quality elements, while limiting the computational time.

The density-based methods are strong regarding the ability to change topology and change the design dramatically, due to the design sensitivities being distributed over a large part of the domain. The cost of introducing the design is relatively low, as only an extra term needs to be integrated, with no special interface treatment being necessary. However, there are problems such as choosing proper interpolation schemes for material properties and, in the case of fluids, a large enough penalisation of the flow in solid regions. The velocity and pressure fields are present in the entire domain, both solid and fluid regions, which may cause spurious flows and leaking pressure fields, if not penalised sufficiently.

For the surface-capturing methods, the well defined and crisp interface makes it easy to introduce interface couplings between different physics, e.g., for fluid–structure interaction problems. As for any level set method, due to the nature of the method, the design sensitivities are located only at the interface. This means that design changes can only propagate from the interface and no new holes appear automatically. This is often relieved by using an initial design with many holes or by introduction of a hole nucleation scheme, e.g., using topological derivatives.

The above methods will not be described in detail, but the readers are referred to the descriptions in the individual papers of the review and the general overview in the review papers by Sigmund and Maute [2] and van Dijk et al. [6].

1.2. Layout of Paper

The included papers are divided into different subsets based on the number and complexity of the physics involved. The layout of this paper is accordingly divided into sections.

The literature review is presented in Section 2. Section 2.1 covers pure fluid flow problems divided into steady laminar flow (Section 2.1.1), unsteady flow (Section 2.1.2), turbulent flow (Section 2.1.3) and non-Newtonian fluids (Section 2.1.4). Section 2.2 considers species transport problems. Section 2.3 deals with conjugate heat transfer problems, where the thermal field is modelled in both solid and fluid domains, divided according to the type of cooling into forced convection (Section 2.3.1) and natural convection (Section 2.3.2). Section 2.5 considers both material microstructures (Section 2.5.1), where effective material parameters are optimised, and porous media (Section 2.5.2), where homogenised properties are used to optimise a macroscale material distribution. Section 2.4 covers fluid–structure interaction (FSI) problems with the fluid flow loading a mechanical structure.

After the literature review, Section 3 performs a quantitative analysis of the included papers and Section 4 presents recommendations for future focus areas for research within topology optimisation of fluid-based problems. Finally, Section 5 briefly concludes the review paper.

2. Literature Review

In the following, the papers are grouped based on their most advanced example, if the work covers both simple and extended applications. Furthermore, the papers are presented in chronological order based on the date that the papers were available online, as this gives a better representation of the order than official date of the final issue, since that may well trail the online publication significantly and differently from paper to paper.

2.1. Fluid Flow

This section covers the majority of the papers included in the review paper, namely pure fluid flow problems. The section is divided into subsections covering steady laminar flow, unsteady flow, turbulent flow, microstructure and porous media.

2.1.1. Steady Laminar Flow

Borrvall and Petersson [7] published the seminal work on fluid topology optimisation in 2003. They presented an in-depth mathematical basis for topology optimisation of Stokes flow. The design parametrisation is based on lubrication theory, leveraging the frictional resistance between parallel plates. Solid domains are approximated by areas with vanishing channel height. By designing the spatially-varying channel height, it is possible to achieve fluid topologies that dictate where flow channels minimising the dissipated energy are placed. This parametrisation was extended to

Navier–Stokes flow by Gersborg-Hansen et al. [8] in 2005. Both sets of authors note the similarity of the obtained equations with that of a Brinkman-type model of Darcy's law for flow through a porous medium. Here, solid domains are approximated by areas with a very low permeability. When treating two-dimensional problems of a finite depth, it makes sense to use the lubrication theory approach, since this ensures the out-of-plane viscous resistance due to finite channel width being taken into account in the fluid parts of the domain. However, for three-dimensional problems, the lubrication theory approach loses its physical meaning, whereas the porous media approach carries over without any issues.

Evgrafov [9] investigated the limits of porous materials in the topology optimisation of Stokes flow using mathematical analysis, complementing the analysis presented originally by Borrvall and Petersson [7]. Olesen et al. [10] presented a high-level programming-language implementation of topology optimisation for steady-state Navier–Stokes flow using the fictitious porous media approach. Guest and Prévost [11] took a different approach to the previous work and modelled the solid region as areas with Darcy flow of low permeability surrounded by areas of Stokes flow using an interpolated Darcy–Stokes finite element. Evgrafov [12] investigated the theoretical foundation and practical stability of the penalised Navier–Stokes equations going to the limit of infinite impermeability, showing that the problem is ill-posed for increasing impermeability of solid regions and that slight compressibility and filters can ensure solutions. Wiker et al. [13] treated problems with separate regions of Stokes and Darcy flow, similar to Guest and Prévost [11], but with a finite impermeability in order to simulate actual flow in porous media for a mass flow distribution problem. Pingen et al. [14] used the Lattice Boltzmann method (LBM) as an approximation of Navier–Stokes flow. Their work included the first three-dimensional result on a very coarse discretisation. Aage et al. [15] were the first to treat truly three-dimensional problems using shared-memory parallelisation, allowing them to optimise large scale Stokes flow problems. Bruns [16] highlighted the similarity of the previous approaches to that of a penalty formulation of imposing no-flow boundary conditions. Specifically, a volumetric penalty term is used to impose no-flow inside solid regions. This interpretation has become widespread through the years, as the physical relevance of the design parametrisation has lost interest and the strictly numerical view has gained popularity. Duan et al. [17–19] presented the first application of a variational level set method to fluid topology optimisation, producing similar designs to those previously obtained using density-based methods. Evgrafov et al. [20] performed a theoretical investigation into the use of kinetic theory to approximate Navier–Stokes for fluid topology optimisation. Othmer [21] derived a continuous adjoint formulation for both topology and shape optimisation of Navier–Stokes flow for implementation into the finite volume solver OpenFOAM. Although results for turbulent flow are presented, the approach is herein not considered "fully turbulent" due to the assumption of "frozen turbulence" (not taking the turbulence variables into account in the sensitivity analysis). Pingen et al. [22] discuss efficient methods for computation of the design sensitivities for LBM based methods and apply topology optimisation to a Tesla-type valve design problem. Zhou and Li [23] presented a variational level set method for Navier–Stokes flow excluding elements outside of the fluid region in order to increase accuracy of the no-slip boundary condition, but increasing book-keeping through updating the fluid mesh every design iteration. Pingen et al. [24] formulated a parametric level set approach using LBM to model the fluid flow. In contrast to the previous variational level set methods, gradient-based mathematical programming is used to update the level set function rather than using the traditional Hamilton–Jacobi equation. Similarly to Zhou and Li [23], Challis and Guest [25] proposed a variational level set method where only discrete fluid areas occur. This removes the need for interpolation schemes, and, by discarding the degrees-of-freedom in the solid, they are able to solve large three-dimensional problems at a reduced cost. Kreissl et al. [26] presented a generalised shape optimisation approach using an explicit level set formulation, where the nodal values of the level set are explicitly varied using a mathematical programming approach in contrast to variational [17–19,23,25] and parametric [24] level set formulations. Furthermore, they use a geometric boundary representation that enforces no-slip conditions along the fluid–solid interface using the LBM. Liu et al. [27] optimised fluid distributors with multiple flow rate equality constraints. Okkels et al. [28] elaborated on the lubrication theory approach to design the height profile of the inlet of a bio-reactor minimising the pressure drop while maintaining a uniform flow through the reactor. While most of the previous works treated energy dissipation or pressure drop, Kondoh et al. [29] presented a density-based topology optimisation formulation for drag minimisation and lift maximisation of bodies using body force integration. Building on their previous work, Kreissl and Maute [30] developed an explicit level set method accurately capturing the boundary using the extended finite element method (X-FEM). Deng et al. [31] presented a density-based method for both steady and unsteady flows driven by gravitational, centrifugal and Coriolis body forces. Deng et al. [32] also proposed an implicit variational level set method for steady Navier–Stokes flow with body forces.

The work by Aage and Lazarov [33] is not focused on fluids, but rather presents a parallel framework for large-scale three-dimensional topology optimisation. However, they use a Stokes flow driven manifold distributor as an application using 3.35 million DOFs for the fluid problem. Evgrafov [34] presented a state space Newton's method for minimum dissipated energy of Stokes flow, which outperforms the traditional first order nested approaches significantly for specific problems. Romero and Silva [35] applied a density-based approach to the design of rotating laminar rotors by including the centrifugal and Coriolis forces. Yaji et al. [36] formulated a level set based topology optimisation of steady flow using the LBM and a continuous adjoint sensitivity analysis based on the Boltzmann equation. Liu et al. [37] presented a density-based method based on the LBM using a discrete adjoint sensitivity analysis posed in the moment space. Yonekura and Kanno [38] suggested to use transient information for steady state optimisation using the LBM by modifying the design during a single transient solve. Liu et al. [39] proposed a re-initialisation method for level set based optimisation using a regular mesh for the level set equation and an adaptive mesh for the fluid analysis. Duan et al. [40] presented an adaptive mesh method for density-based optimisation using an optimality criteria (OC) algorithm. Extending his previous work, Evgrafov [41] presented a Chebyshev method for topology optimisation of Stokes flow achieving locally cubic convergence for specific problems. Garcke and Hecht [42] performed an in-depth mathematical analysis of a phase field approach to topology optimisation of Stokes flow. Lin et al. [43] applied a density-based approach to the design of fixed-geometry fluid diodes, manufacturing the designs and verifying their performance using an experimental setup. Building on their previous formulation and analysis, Garcke et al. [44] provided numerical results for Stokes flow using their phase field approach and adaptive mesh refinement. Duan et al. [45] presented an Ersatz material based implicit level set method showing clear connections to density-based methods through the material distribution. Sá et al. [46] applied the concept of topological derivatives to fluid flow channel design.

Yoshimura et al. [47] proposed a gradient-free approach using a genetic algorithm to update a very coarse design using a Kriging surrogate model coupled to an immersed method known as the Building-Cube Method. Pereira et al. [48] applied a density-based approach to Stokes flow using polygonal elements and supplying a freely available code for fluid topology optimisation. Duan et al. [49] presented an adaptive mesh method to an Ersatz-material level set based approach for Navier–Stokes flow. Kubo et al. [50] used a variational Ersatz-material level set method to design manifolds for flow uniformity in microchannel reactors. Jang and Lee [51] maximise the acoustic transmission loss in a chamber muffler with a constraint on the reverse-flow power dissipation modelled using the Navier–Stokes equations. Koch et al. [52] proposed a method for automatic conversion from two-dimensional Ersatz-based level set topology optimisation to NURBS-based shape optimisation. Sato et al. [53] applied density-based topology optimisation to the design of no-moving parts fluid valves using a Pareto front exploration method. Sá et al. [54] further extended the work in [46] to rotating domains. Yonekura and Kanno [55] extended their previous approach for updating the design during the computation of an unsteady flow field to a level set based method. Dai et al. [56] presented a piecewise constant Ersatz-material level set method, which essentially reduces it to a density-based method. Shen et al. [57] formulated a three-phase interpolation model in Darcy–Stokes flow for optimising fluid devices with Stokes flow, Darcy flow and solid domains. Deng et al. [58] used an Ersatz-material level set method to optimise two-phase flow, using with a phase field approach for the fluid–fluid interface. Garcke et al. [59] extended their phase field approach to Navier–Stokes flow, presenting both in-depth mathematical analysis and optimisation results for minimum drag and maximum lift problems. Alonso et al. [60] used a density-based approach to design laminar rotating swirl flow devices using a rotating axisymmetric model.

Jensen [61] proposed to use anisotropic mesh adaptation for density-based optimisation of Stokes flow, demonstrating that this allows for efficiently resolving the physical length scale related to very high Brinkman penalisation terms. Sá et al. [62] applied density-based optimisation to the design of a small scale rotary pump considering both energy dissipation and vorticity with experimental verification of the design performance. Zhou et al. [63] presented an integrated shape morphing and topology optimisation approach based on a mesh handling methodology called a deformable simplicial complex (DSC). Shin et al. [64] used a 2D axisymmetric model at moderate Reynolds numbers to optimise a vortex-type fluid diode that is postprocessed and verified by simulation under actual flow conditions and turbulence. Yonekura and Kanno [65] proposed a heuristic approximation of the Hessian matrix for fast density-based optimisation using the LBM. Behrou et al. [66] presented a methodology for adaptive explicit no-slip boundary conditions with a density-based method for laminar flow problems with mass flow constraints, adaptively removing elements in the solid regions. Alonso et al. [67] applied density-based optimisation to the design of Tesla-type centrifugal pumps without blades. Lim et al. [68] applied a density-based method to the design of vortex-type passive fluidic diode valves for nuclear applications. Sato et al. [69] presented a topology optimisation method for rarefied gas flow problems covering both gas and solid domains using the LBM. Gaymann et al. [70] applied a density-based method to the design of fluidic diode valves in both two- and three-dimensions at medium-to-high Reynolds numbers. Gaymann and Montomoli [71] applied deep neural networks and Monte Carlo Tree search to an absurdly coarse design grid.

2.1.2. Unsteady Flow

All of the previous works consider steady-state flow problems. However, not all fluid flows develop a steady state and there are situations where the unsteady motion can be exploited by the design. This could either be due to transients in an onset flow or due to vortex shedding from an obstacle. It should be noted that some discretisation methods, e.g., lattice Boltzmann methods, are of a transient nature; however, the papers are categorised based on the use of a time-dependent objective functional and sensitivity analysis. Thus, if only the final state solution is used to update the design, it is considered steady state and not included in this section.

Kreissl et al. [72] presented the first work treating density-based topology optimisation for unsteady flow, using a discrete transient adjoint formulation and applied to the design of a diffuser and an oscillating flow manifold. They highlighted some difficulties encountered using a standard density approach combined with a stabilised finite element formulation. A few months later, Deng et al. [73] also presented density-based topology optimisation for unsteady flow, but using a continuous transient adjoint formulation. They optimised a wide arrangement of problems including oscillating manifold flows. Deng et al. [31] extended their work to both steady and unsteady flows driven by gravitational, centrifugal and Coriolis body forces. Abdelwahed and Hassine [74] introduced analysis and application of the topological gradient method for non-stationary fluid flows. Nørgaard et al. [75] presented topology optimisation of what can be considered the first truly unsteady flow problems, considering oscillating flow over multiple periods for both obstacle reconstruction and design of an oscillating pump using the LBM. Villanueva and Maute [76] formulated a CutFEM discretised explicit level set method to the design of two- and three-dimensional flow for both steady-state and fully transient problems. Although the use of a surface-capturing scheme, they show that there is still a penalty- and mesh-dependent mass loss through the interface. Chen et al. [77] presented a local-in-time approximate discrete adjoint sensitivity analysis for unsteady flow using the LBM. They approximate the true time-dependent adjoint equations by splitting the time series into subintervals, rather than the full time series forward and then the full time series backwards. This significantly reduces the storage requirement but introduces an approximation error. Nørgaard et al. [78] discussed applications of automatic differentiation for topology optimisation, demonstrating their approach on an unsteady oscillating pressure pump. This was obtained using LBM and the design seems to rely on the slightly compressible behaviour of the fluid model. Sasaki et al. [79] presented fluid topology optimisation using a particle method for the first time. Specifically, they use the transient moving particle semi-implicit (MPS) method allowing them to treat free surface fluid flows without explicit surface tracking.

2.1.3. Turbulent Flow

All of the above fluid flow papers assume laminar fluid flow, whereas turbulent flow has only been treated in a few publications. Turbulence is inherently time-dependent, but current works on topology optimisation of turbulent flow restrict themselves to the steady-state time-averaged approximation of turbulence, namely the Reynolds-Averaged Navier–Stokes (RANS) equations.

In the work by Othmer [21], turbulence is in the model, but the influence on the design sensitivities is neglected. Kontoleontos et al. [80] presented the first work on topology optimisation of turbulent flow, including the Spalart–Allmaras turbulence model in their continuous adjoint sensitivity analysis. For a shape optimisation example, they showed that the typical "frozen turbulence" assumption produces sensitivities of the incorrect sign in some cases. Yoon [81] presented a discrete adjoint approach to density-based optimisation of turbulent flow problems using the Spalart–Allmaras turbulence model and a modified wall equation. However, the meshes used are much too coarse to capture turbulence properly. Dilgen et al. [82] demonstrated the application of automatic differentiation for obtaining exact sensitivities for density-based topology optimisation of large scale two- and three-dimensional turbulent flow problems, using the one-equation Spalart–Allmaras model and the two-equation k-ω model. As Kontoleontos et al. [80] did for shape optimisation, they demonstrated that the "frozen turbulence" assumption gives inexact sensitivities for topology optimisation, even with the incorrect sign for some cases. Yoon [83] used a $k - \epsilon$ model, analogous to the model applied [82], to design 2D flow components minimising turbulent energy i.e., minimising noise.

2.1.4. Non-Newtonian Fluids

In the preceding works, the fluids are all assumed to be Newtonian. However, treating more sophisticated fluids, including e.g., long polymer chains or blood cells, calls for implementation of non-Newtonian fluids, which have nonlinear behaviour of the viscosity. There are a wide variety of models that can be applied and a few have been implemented for use in a topology optimisation context.

Pingen and Maute [84] applied topology optimisation to non-Newtonian flows for the first time, using a density-based LBM formulation and a Carreau–Yasuda model for shear-thinning fluids. Ejlebjerg Jensen et al. [85] optimised viscoelastic rectifiers using a non-Newtonian fluid model based on dumbbells in a Newtonian solvent, which introduced a memory in the fluid. Jensen et al. [86] considered the bi-stability behaviour for a crossing between two viscoelastic fluids. Hyun et al. [87] suggested a density-based formulation for minimising wall shear stress by considering shear thinning non-Newtonian effects. Zhang and Liu [88] applied a level set based approach to minimise flow shear stress in arterial bypass graft designs, where the blood flow is modelled using a steady non-Newtonian modified Cross model. Zhang et al. [89] used an explicit boundary-tracking level set method with remeshing to optimise micropumps for non-Newtonian power-law fluids. Romero and Silva [90] extended their previous work [35] to cover non-Newtonian fluids and compare it to the Newtonian case. Dong and Liu [91] proposed a bi-objective formulation for the design of asymmetrical fixed-geometry microvalves for non-Newtonian flow.

2.2. Species Transport

In this section, the included papers are focused on a transport of matter or species due to the presence of a fluid. The transported matter does not necessarily need to be modelled itself, as long as the objective of the optimisation is related to the transport.

Okkels and Bruus [92] coupled a convection–reaction–diffusion equation to the fluid flow to model catalytic reactions, distributing the porous catalytic support to maximise the mean reaction rate of the microreactor. Andreasen et al. [93] used a convection–diffusion equation to model and optimise microfluidic mixers, in which well-known design elements such as herring-bones and slanted grooves appeared automatically. Gregersen et al. [94] applied topology optimisation to an electrokinetic model in order to maximise the net induced electroosmotic flow rate. Schäpper et al. [95] used more advanced reaction-kinetics using multiple convection-diffusion type equations to optimise microbioreactors. Kim and Sun [96] optimised the gas distribution channels in automotive fuel cells. Makhija et al. [97] optimised a passive micromixer using a porosity model for LBM. Deng et al. [98] used a physical model similar to [93] but omits the pressure constraint and use a quasi-Newton approach optimised three-dimensional and extruded two-dimensional microfluidic mixers. Makhija and Maute [99] introduced an explicit level set optimisation methodology using an X-FEM-based hydrodynamic Boltzmann model including transport. The ability to eliminate the spurious diffusion in void areas, especially dubious when modeling species concentration, is highlighted. Oh et al. [100] used the Navier–Stokes and a convection-diffusion equation to model and optimise the osmotic permeate flux over a membrane wall. Chen and Li [101] optimised micromixers under the assumption that reverse flow structures [8] inserted in a microchannel increases the mixing. Hyun et al. [102] designed repeating units for sorting particles using principles in deterministic lateral displacement. Andreasen [103] used a density-based framework to design dosing units of a secondary fluid utilising the inertia of the driving fluid. Yaji et al. [104] presented an optimisation of vanadium redox flow batteries by including a reaction term depending on the local flow speed and concentration level in a two-dimensional setting. Guo et al. [105] presented a methodology to model and optimise pure convection-dominated transport using a Lagrangian mapping method. Only the Navier–Stokes equations are approximated by FEM, while the Lagrangian transport is modelled cross-section-wise. Behrou et al. [106] presented a density-based approach for the design of proton exchange membrane fuel cells using a depth-averaged two-dimensional approximation of reactive porous media flow. Chen et al. [107] extended their previous work [104] to three-dimensional problems. Dugast et al. [108] used a level set method and the LBM to maximise the reaction in a square reactor and investigated the problem for a range of flow situations.

2.3. Conjugate Heat Transfer

Conjugate heat transfer is when the coupled heat transfer between a solid and the surrounding fluid is considered, with the temperature field of both of interest. Thus, in order to model conjugate heat transfer, it is necessary to build the thermal transport on top of the fluid flow model. Conjugate heat transfer is generally divided into groups based on the heat transfer mechanism in the fluid.

Figure 2 shows the three main heat transfer mechanisms: forced convection, where the flow is actively driven by a pump, fan or pressure-gradient; natural convection, where the flow happens passively from the natural density variations due to temperature differences; and diffusion where heat is transferred through a stagnant fluid through diffusion. Only the first two are considered in this review, since they include fluid motion modelled through fluid flow equations.

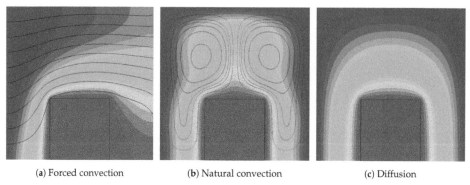

(a) Forced convection (b) Natural convection (c) Diffusion

Figure 2. Illustration of a metallic block subjected to different heat transfer mechanism in the surrounding fluid. (**a**) shows forced convection with a cold flow entering at the left-hand side; (**b**,**c**) show natural convection and pure diffusion, respectively, due to cold upper and side walls. Reproduced with permission from Alexandersen et al. [109].

2.3.1. Forced Convection

Dede [110] and Yoon [111] presented the first works on topology optimisation of forced convection at almost the same time. Dede [110] used the commercial finite element analysis (FEA) software COMSOL to optimise both conduction and conjugate heat transfer problems. Yoon [111] presented a two-dimensional formulation, treating heat sink problems, as well as flow focusing in order to cool specific points. Thereafter, Dede [112] applied topology optimisation to design multipass branching microchannel heat sinks for electronics cooling. McConnell and Pingen [113] presented a two-layer pseudo-3D topology optimisation formulation based on the lattice Boltzmann method (LBM). Kontoleontos et al. [80] presented a continuous adjoint formulation for fluid heat transfer using the Spalart–Allmaras turbulence model and the impermeability directly as the design variable. However, the presented examples are not true conjugate heat transfer problems, since the solid temperature is predefined and enforced through a penalty approach, rather than modelling the solid temperature alongside the fluid temperature. Matsumori et al. [114] presented topology optimisation for forced convection heat sinks under constant input power. They interpolated the heat source to only be active in the solid and investigated both temperature-independent and -dependent sources. Marck et al. [115] investigated a multiobjective optimisation problem considering both fluid and thermal objectives using a finite volume-based discrete adjoint approach. Koga et al. [116] presented the development of an active cooling heat sink device using topology optimisation, which was manufactured and experimentally tested. However, they used a two-dimensional Stokes flow model to optimise for a three-dimensional turbulent application. In 2015, Yaji et al. [117] published three-dimensional results for forced liquid-cooled heat sinks using an Ersatz-material level set approach. Next, Yaji et al. [118] presented a topology optimisation method using the Lattice Boltzmann Method (LBM) incorporating a special sensitivity analysis based on the discrete velocity Boltzmann equation. Łaniewski Wołłk and Rokicki [119] treated large three-dimensional problems using a discrete adjoint formulation for the LBM implemented for multi-GPU architectures. Qian and Dede [120] introduced a constraint on the tangential thermal gradient around discrete heat sources with the goal of reducing thermal stress due to non-uniform expansion. Yoshimura et al. [47] proposed a gradient-free approach using a genetic algorithm and a Kriging surrogate model coupled to an immersed method known as the Building-Cube Method. Haertel and Nellis [121] developed a plane two-dimensional fully-developed flow model for topology optimisation of air-cooled heat sinks. Pietropaoli et al. [122] used the impermeability as the design variable to optimise internal channels.

In 2018, Zhao et al. [123] used a Darcy flow model for topology optimisation of cooling channels. Qian et al. [124] optimised active cooling flow channels for cooling an active phased array antenna with many discrete heat sources. Sato et al. [125] used an adaptive weighting scheme for the multiobjective topology optimisation of active heat sinks. Yaji et al. [126] applied a local-in-time approximate transient adjoint method for topology optimisation of large scale problems with oscillating inlet flow. Haertel et al. [127] presented a pseudo-3D model for extruded forced convection heat sinks, actually considering the chip temperature by coupling a thermofluid design layer to a conductive base plate layer. Almost simultaneously, Zeng et al. [128] published a similar two-layer model for an air-cooled mini-channel heat sink, where the connection between the layers is tuned using full three-dimensional simulations of a reference heat sink design. Furthermore, Zeng et al. [128] manufactured and experimentally validated the performance of their optimised designs. Dilgen et al. [129] presented the first full conjugate heat transfer model for density-based topology optimisation of turbulent systems. In contrast to Kontoleontos et al. [80], the temperature field of the solid is modelled, thus rendering it true conjugate heat transfer. Furthermore, Dilgen et al. [129] treat large-scale three-dimensional problems comparing their thermal performance to equivalent two-dimensional designs. Ramalingom et al. [130] proposed a sigmoid interpolation function for mixed convection problems. Dugast et al. [131] applied a level set based approach in combination with the LBM to a variety of thermal control problems. Santhanakrishnan et al. [132] performed a comparison of density-based and Ersats-material level set topology optimisation for three-dimensional heat sink design using the commerical FEA software COMSOL. However, the designs, both density and level set based, show clear signs of being unconverged with unphysical designs and, thus, the study must be rendered inconclusive. Sun et al. [133] used density-based topology optimisation to generate guiding channels for an enhanced air-side heat transfer geometry in fin and tube heat exchangers. Lv and Liu [134] applied a density-based method to the design of a bifurcation micro-channel heat sink, comparing them to reference designs.

In 2019, Pietropaoli et al. [135] extended their previous work to three-dimensional internal coolant systems. Makhija and Beran [136] presented a concurrent optimisation method using a shape parametrisation for the external shape and a density-based parametrisation for the internal geometry. Subramaniam et al. [137] investigated the inherent competition between heat transfer and pressure drop. Yu et al. [138] applied a geometry projection method called moving morphable components (MMC) to the design of two-dimensional problems allowing for explicit feature size contol. Zhang and Gao [139] presented a density-based approach for optimising non-Newtonian fluid based thermal devices. Kobayashi et al. [140] used topology optimisation to design extruded winglets for fin-and-tube heat exchangers. Zeng and Lee [141] extended their previous work to the design of liquid-cooled microchannel heat sinks with in-depth numerical and experimental investigations. Jahan et al. [142] designed conformal cooling channels for plastic injection molds using a two-dimensional simplification. Yan et al. [143] developed a two-layer plane model based on analytical derivations and assumptions of the out-of-plane distribution for optimising microchannel heat sinks. Tawk et al. [144] proposed a density-based approach for optimising heat exchangers with two seperate fluids and a solid. Lundgaard et al. [145] presented a density-based methodology for distributing sand and rocks in thermal energy storage systems modelled by a transient Darcy's law coupled to heat transfer. Li et al. [146] applied a multi-objective density-based method to the design of liquid-cooled heat sinks, presented both extensive numerical and experimental comparisons to reference designs. Dong and Liu [147] applied topology optimisation to air-cooled microchannel heat sinks with discrete heat sources. Yaji et al. [148] suggested a multifidelity approximation framework to optimise turbulent heat transfer problems using a low-fidelity laminar flow model as the driver. Hu et al. [149] applied a density-based approach to optimisation of a microchannel heatsink with an in-depth comparison to a reference design with straight channels.

2.3.2. Natural Convection

Alexandersen et al. [109] presented the first work on topology optimisation of natural convection problems, using a density-based approach for optimising both heat sinks and buoyancy-driven micropumps. On the contrary, Coffin and Maute [150] used an explicit level set method combined with the extended finite element method (X-FEM) for both steady-state and transient natural convection cooling problems. Alexandersen et al. [151] extended their initial paper to large-scale three-dimensional heat sink problems using a parallel framework allowing for the optimisation of problems with up to 330 million DOFs. Pizzolato et al. [152] applied topology optimisation to the design of fins in shell-and-tube latent heat thermal energy storage, including the temperature-dependent latent heat coupled with natural convection using a time-dependent formulation. Alexandersen et al. [153] applied their previously developed framework to optimise the design of passive coolers for light-emitting diode (LED) lamps showing superior performance compared to reference lattice and pin fin designs. Ramalingom et al. [130] proposed a sigmoid interpolation function for mixed convection problems. Lazarov et al. [154] performed an experimental validation of the optimised designs from Alexandersen et al. [153] using additive manufacturing in aluminium, showing good agreement with numerical results and highlighting the superiority of topology-optimised designs. Lei et al. [155] continued this work and used investment casting to experimentally investigate a larger array of heat sink designs comparing them to optimised pin fin designs. Saglietti et al. [156] presented topology optimisation of heat sinks in a square differentially heated cavity using a spectral element method. In order to reduce the computational cost, Asmussen et al. [157] suggested an approximate flow model to that originally presented by Alexandersen et al. [109], by neglecting intertia and viscous boundary layers. Pizzolato et al. [158] extended their previous work to maximise the performance of multi-tube latent heat thermal energy storage systems, investigating many different working conditions. Ramalingom et al. [159] applied their previous method to multi-objective optimisation of mixed and natural convection in a asymmetrically-heated vertical channel. Pollini et al. [160] extended the work of Asmussen et al. [157] to large-scale three-dimensional problems, producing results comparable to those of Alexandersen et al. [151] with a computational time reduction of 80–95% in terms of core-hours.

2.4. Fluid–Structure Interaction

In this section, the advances within topology optimisation of fluid–structure interaction (FSI) problems are discussed.

Figure 3 shows the types of design modifications possible for FSI problems. Dry optimisation only changes the internal structure, keeping the solid–fluid interface constant. Wet optimisation modifies the solid–fluid interface and topology. Since this review paper is focused on the optimisation of the fluid flow, only works where the wet surface and topology is allowed to change significantly are included (wet and wet+dry). A whole range of works describe the topology optimisation of structural parts subjected to fluid loads, where the deformation of the structure may or may not be taken into account when computing the fluid induced loads. However, because they do not modify the wet surface and topology, they are not considered herein.

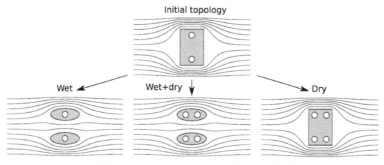

Figure 3. Description of different degrees of design modification for fluid–structure interaction (FSI) problems.

Yoon [161] can be considered the seminal paper on topology optimisation for FSI. A unified density-based formulation of the elastic Navier–Cauchy equations assuming small strains and the incompressible Navier–Stokes equation is obtained by converting the interface condition to a volumetric integral representation, previously used for acoustic–structure interaction [162]. The fluid stress in the interaction is slightly simplified and a pressure filter function determines where the fluid pressure applies. The fluid problem is solved in the deformed mesh and a full coupling is modelled; however, the obtained deformations of the solid domain are extremely small and the two-way coupling is not really active. Another approach was taken by Kreissl et al. [163], where micro-fluidic devices are optimised subject to external mechanical actuation. A one-way coupling from structure to fluid is used to deform the fluid domain. The backward fluid-structural coupling is assumed negligible and hence ignored. Yoon [164] extends the previous work [161] to cover electro-fluid-thermal-compliant actuators, including two additional physical fields, electric and thermal, in the coupling. Planar multiphysics MEMS devices are optimised with electrical and thermal response being computed in the reference mesh. Subsequently, Yoon extended the framework to minimise the structural mass subject to stress constraints [165], and also applied the framework to the optimisation of a compliant flapper valve [166].

Jenkins and Maute [167] presented a coupled level set based framework utilising X-FEM and deformed meshes, demonstrating generalised shape optimisation of a bio-prosthetic heart valve and topology optimisation of the wall example of Yoon [161]. Munk et al. [168] present a simplified model for designing baffle plates, with a one-way pressure coupling omitting fluid shear stress. The fluid is modelled with LBM and the loads are mapped to the structural model in the reference configuration. The Bi-directional Evolutionary Structural Optimisation (BESO) method is used in a soft-kill version, but the sensitivities of changing the fluid flow are neglected. Similarly, Picelli et al. [169] also neglect the sensitivities of changing the fluid flow when updating the wet surface when applying a hard-kill BESO method to design various FSI problems.

Yoon [170] presented an extension of previous work [165], where a material failure criteria is applied to design the material distribution. Lundgaard et al. [171] revisited the unified density-based formulation of Yoon [161], however, solving the fluid in the reference mesh under the assumption of small deformations. Multiple objective functions and design problems are reviewed and thorough discussions of current limitations, artefacts and future extensions for density-based topology optimisation of FSI problems are given. Munk et al. [172] compared the previous formulation [168] to level set and density-based methods for the case of minimising the compliance of a fluid-loaded baffle plate. Subsequently, Munk et al. [173] ported the work to graphics processing unit (GPU) architecture in order to reduce the high computational time for the LBM model. Feppon et al. [174] used a level set-based framework to explicitly track and advance the interface using the Hamilton-Jacobi equation. The meshes are iteratively updated based on the convected level set by local operations, with the physics being weakly coupled, modeled in referenced configuration and solved using a staggered procedure.

2.5. Microstructure and Porous Media

In relation to the origin of topology optimisation, namely the homogenisation approach, there are a range of studies that consider the optimisation of material microstructures. Typically a unit-cell, or representative volume element (RVE), is subjected to periodic boundary conditions and the effective parameters are obtained by imposing a set of volumetric loads. This approach can be utilised in an inverse manner to optimise the material design and the corresponding effective parameters. For solids, this is related to the effective stiffness tensor, while for fluids this is naturally related to the permeability. For fluid–structure interaction in a porous medium, a pressure coupling term can also be obtained by homogenisation.

2.5.1. Material Microstructures

Guest and Prévost [175] maximised the permeability of a porous material microstructures using a Darcy–Stokes interpolation [11] subject to isotropic symmetry constraints. The work is extended in Guest and Prévost [176] to optimise microstructures for combined maximum stiffness and permeability. Bones and tissue contain porous materials and Hollister and Lin [177] optimised tissue engineering scaffolds using a hybrid stiffness and permeability optimisation routine. Xu and Cheng [178] proposed a multiscale optimisation problem, where the macroscopic elastic compliance is minimised subject to a flow constraint ensuring a permeable microstructure. Physically related to this, Andreasen and Sigmund [179] optimised the microstructure of a poroelastic material for maximum poroelastic coupling during pressurisation subject to permeability constraints. Chen et al. [180] studied the optimisation of bio-scaffolds using homogenisation for tissue regeneration including permeability considerations. Chen et al. [181] extended the work to consider shear induced wall erosion. Goncalves Coelho et al. [182] introduced permeability constraints in an extensive multiscale optimisation framework for the multiscale topology optimisation of trabecular bone. For most material properties, certain bounds apply in property space and Challis et al. [183] investigated the cross-property bounds between stiffness and permeability by exploiting the Pareto-front using a level-set based approach [25].

2.5.2. Porous Media

In this subsection, works where the final design is supposed to be a porous structure i.e., intermediate design variables, are reviewed. This can be in terms of multiscale problems obtained e.g., by two-scale asymptotic expansion. A different take on FSI problems was presented by Andreasen and Sigmund [184] for optimisation of material design for poroelastic actuators and in Andreasen and Sigmund [185] for impact energy absorption in porous structures. Furthermore, in the context of macroscale problems, Youssef et al. [186] optimised a porous scaffold with macroscale flow channels to control the internal shear stress in a bioreactor. Ha et al. [187] used the Darcy–Stokes interpolation method [175] to maximise the permeability of three-dimensional woven materials. A multimaterial approach is taken by Wein et al. [188], where a highly nonlinear saturated porous model with multiple materials is applied to design diapers that quickly transports fluid away from the surface to capture it in the interior. Takezawa et al. [189] used a Brinkman–Forchheimer macro model to optimise material microstructures for minimum flow resistance considering the trade-off between permeability and form-drag. Lurie et al. [190] optimised the distribution of the wick (porous media used to transport condensate to the evaporator due to capillary effects) in a heat pipe. Takezawa et al. [191] applied a multiscale method to the thermofluid problem of metal printed lattice design. Effective parameters for permeability, form drag and conductivity are obtained for a generic orthogonal truss microstructure and used in a macroscopic material distribution method based on the Brinkman–Forchheimer and a convection–diffusion equation. Takezawa et al. [192] later extended this to the fluid-thermo-elastic problem of metal printed heat sinks.

3. Quantitative Analysis

In this section, a quantitative study of the referenced papers is carried out. In some cases, the method details might be unclear or not mentioned, excluding the paper from the statistics.

3.1. Total Publications

Figure 4 shows the number of papers published per year and the total accumulated number of publications over time since the inaugural paper by Borrvall and Petersson [7] in 2003. The year of publication is here taken as the year of the final journal issue.

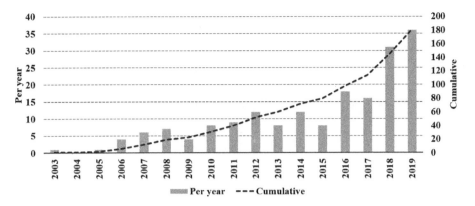

Figure 4. Number of papers published per year and total accumulated publications over time.

From 2003 to 2015, a slow increase in the number of papers per year is observed, reaching an average of 10 per year for the period 2012–2015. In 2016 and 2017, the number of papers per year almost doubled to 18 and 16, respectively, bringing the total number of publications to 100 after 2016. In 2018 and 2019, the number of papers per year again almost doubled to 31 and 36, respectively. The total number of papers by the end of 2019 reached 182. This approximate doubling behaviour shows itself as an exponential-like increase in the number of total papers. In 2020, covering only the month of January, there have been five publications so far. This amounts to a total of 186 papers covered by this review.

3.2. Design Representations

As discussed in Section 1, several different design representations exist for topology optimisation. The papers are grouped into three groups: "density" covering interpolation and homogenisation approaches; "level set" covering Ersatz-material, adapted (here refering to adapted meshes and/or ignoring the solid elements in computations.) and surface-capturing level set approaches; "other" covering anything else, e.g., BESO.

Figure 5 shows how the included papers are distributed among the two main design representations, density-based approaches and level set approaches. Firstly, the use of density-based approaches vastly outnumbers any other methods with 144 papers or 77%. This reflects the general tendency within the topology optimisation community to prefer density-based methods [2,3]. Secondly, the approaches relying on an implicit level set description of the geometry are also numerous at 31 papers or 17%. Of these 31 papers, 11 use an Ersatz-material, 11 use an adapted approach and only nine use a surface-capturing discretisation method. Lastly, the rest of the papers are distributed as follows: 5 using BESO [168,169,172,173,177]; 2 using phase field [42,59]; 1 using a discrete surface representation [63]; 1 using a geometry-projection method [138]; and 2 utilising the topological gradient [46,74].

3.3. Discretisation Methods

For discretising the design and physics, a variety of methods are used in the included papers: finite element method (FEM); finite volume method (FVM); lattice Boltzmann method (LBM) including Boltzmann equation related schemes; particle-based methods (PM).

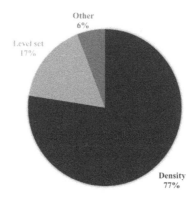

Figure 5. Distribution of papers in overall design representation type.

Figure 6 shows the distribution of papers in these overall methods. It is clear that FEM is the most widely used discretisation method with 134 papers or 76%. The next most used method is LBM at 28 papers [14,20,22,24,26,36–38,44,55,65,69,75,77,78,84,97,108,113,118,119,126,131,163,168,172,173,186] or 16%. PM is the least used method with only a single paper [79]. Surprisingly, FVM is the second least used method with only 12 papers [21,47,52,80,82,115,129,130,137,144,159,188] or 7%, despite the fact that FVM for many years has been the preferred discretisation method for computational fluid dynamics. This can probably be explained by several factors: topology optimisation originates from solid mechanics where FEM is the preferred method; discrete adjoint approaches are easier using FEM than FVM; stabilised FEM has grown to be a mature and accurate method [193,194].

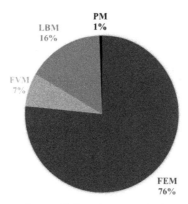

Figure 6. Distribution of papers in overall discretisation method: FEM = finite element methods; FVM = finite volume methods; LBM = lattice Boltzmann methods; PM = particle-based methods.

3.4. Problem Types

The main problem types treated in this review is: pure fluid (PF); species transport (ST); conjugate heat transfer (CHT); fluid–structure interaction (FSI); microstructure and porous media (MP).

Figure 7 shows the distribution of the included papers in these problem types. It is clearly seen that the largest number of papers deal with purely fluid flow problems, namely 82 papers or 44%. The second largest group deals with conjugate heat transfer covering 55 papers or 30%. Species transport covers 19 papers or 10%, with FSI covering 15 papers or 8%, and porous media covering 14 papers or 8%.

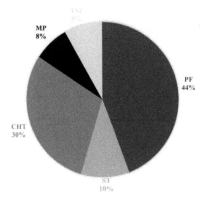

Figure 7. Distribution of papers in overall problem type: PF = pure fluid; ST = species transport; CHT = conjugate heat transfer; FSI = fluid–structure interaction; MP = microstructure and porous media.

3.5. Flow Types

In this review, the fluid model type can be boiled down to four categories: steady-state laminar flow (SS); transient laminar flow (TR); steady-state turbulent flow (TU); and Non-Newtonian fluid (NN).

Figure 8 shows the distribution of papers for fluid model type, both for (a) all papers and (b) papers treating only fluid flow. Analysing all papers, the vast majority use a steady-state laminar flow model with 158 papers or 85% as seen in Figure 8a. Only 13 papers or 7% consider a transient laminar flow model [72–79,145,150,152,158,188], with a meager six papers or 3% treating turbulent flow [21,80–83,129]. In the case of time-dependent problems, this is most likely due to the vast increase in computational cost related to simulation of transient flow problems, where all temporal details must be resolved sufficiently and all temporal solutions saved in memory (or recomputed) for the adjoint solve. Likewise, turbulent flow also carries an increase in computational cost with it, since turbulence models with additional degrees-of-freedom are used and fine meshes are needed to resolve the turbulent boundary layers.

All of the above use a Newtonian fluid model, but nine papers or 5% use a Non-Newtonian model [84–91,139].

Looking only at papers treating fluid flow only, Figure 8b shows that the percentage of the more complex flow models increases. This indicates that more work has been done on treating transient, turbulent and non-Newtonian models for fluid flow only compared to overall for fluid-based problems. This makes sense since pure fluid flow is the obvious place to start working with and tackling the large computational cost associated with the more complex flow models.

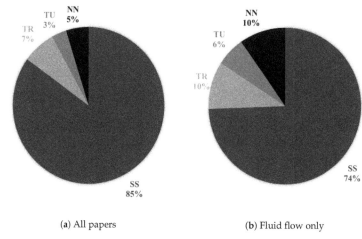

(a) All papers   (b) Fluid flow only

Figure 8. Distribution of papers for fluid model type: SS = steady-state laminar flow; TR = transient laminar flow; TU = turbulent flow; NN = Non-Newtonian fluid.

3.6. Three-Dimensional Problems

A paper is classified as treating three-dimensional problems only if at least one example uses a three-dimensional model in the optimisation process. Therefore, two-dimensional results that are extruded and post-analysed in three dimensions are not included.

Figure 9 shows the distribution of papers treating two-dimensional and three-dimensional problems.

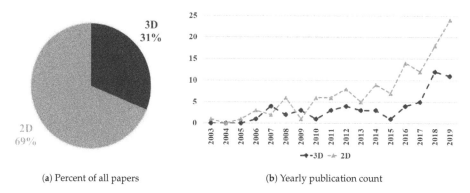

(a) Percent of all papers   (b) Yearly publication count

Figure 9. Distribution of papers for two-dimensional (2D) and three-dimensional (3D) problems.

Figure 9a shows that a significant share of the included papers contain three-dimensional results, namely 31% or 58 out of 186 papers. For the pure fluid flow, there are 15 papers for steady laminar flow [14,15,21,23,25,32–34,36,46,55,56,61,66,70], 3 for unsteady flow [74,76,77], 2 for turbulent flow [80,82] and none for non-Newtonian fluids. For species transport, there are four papers [93,98,105,107]. For conjugate heat transfer, there are eight in forced convection [110,117,119,122,126,129,132,135] and six in natural convection [150,151,153–155,160]. The fluid–structure interaction category counts four papers [164,168,172,173], but it must be noted that, common for all, the three-dimensional design freedom is severely limited, as the design domain is restricted in the third dimension. Finally, a very large share of the three-dimensional results belong in the category of microstructure and porous media problems with 14 papers [175–177,179–184,187–189,191,192]. Since, in material design, it is natural to consider both stiffness and permeability, the need for three-dimensional design freedom is obvious, as two-dimensional models would have either the fluid or the solid phase disconnected.

Figure 9b shows the yearly publication count for two- and three-dimensional problems. It can be seen that they follow the same trend, which is reflected in the fact that the percentage of three-dimensional papers has been close to constant since 2010 at approximately 30%. In January 2020, there has been a single three-dimensional paper [160] out of 5, making it 20% so far.

4. Recommendations

Based on the extensive literature review, analysis of the methods used, as well as the authors' personal experience and opinions, some recommendations are made to the research community in order to help with moving forward.

4.1. Optimisation Methods

Ninety-eight percent of the included papers use gradient-based optimisation approaches, covering amongst others nonlinear programming algorithms, velocity-based level set updates and discrete BESO updates. Most of these use first-order methods, with notable exceptions being the work of Evgrafov [34,41] using higher-order schemes. Only three papers use gradient-free optimisation approaches, consisting of genetic algorithms [47,186] and neural networks [71]. As pointed out by Sigmund et al. [195], gradient-free approaches seldomly make sense for topology optimisation, due to the high dimensional problems for increasing design resolutions. This is perfectly illustrated in the work of Gaymann and Montomoli [71], where the design resolution is absurdly coarse and useless in practise. However, gradient-free approaches can be useful when gradient information is not available, like when using a commercial solver as a black-box, or when dealing with discontinuous functions with non-well-defined gradients. However, even in the case of black-box solvers, gradients can easily be approximated using finite differences at a fraction of the cost of most genetic algorithms. Furthermore, gradient-free methods may have advantages for multi-objective problems, although these can also be included in gradient-based approaches, e.g., [125].

However, gradient-free methods should in general be avoided for topology optimisation of fluid-based problems, and the recommendations of Sigmund [195] should be followed.

4.2. Density-Based Approaches

For density-based approaches, the interpolation of material properties between solid and fluid is of utmost importance to ensure final designs without intermediate design variables. Especially when moving to multiple physics, with an increasing number of material properties, the complexity of choosing the correct form of interpolation increases substantially, see, e.g., [145]. It is often not easy to intuitively choose the various interpolation functions to provide a correct relation between the material properties for intermediate design field values. Thus, it is necessary to either perform analytical derivations (often not possible) or numerical experiments. It is important to investigate the behaviour of the chosen objective functional with respect to the design field to ensure the interpolation functions provides well-scaled and monotonic behaviour [171,196]. Furthermore, it is extremely important to rigorously validate adjoint sensitivities with other methods, such as the complex step method or finite difference approximations as discussed by Lundgaard et al. [145].

Relying on Brinkman penalisation to model an immersed solid geometry in a unified domain has its drawbacks. Due to the nature of the penalisation, there will always exist fluid flow inside the solid. The penalisation factor must be large enough to ensure this flow is negligible inside the solid domain, but small enough to ensure numerical stability of the solution and optimisation algorithms. Generally, this is not observed to be an issue in general, except for a few very specific problems, where pressure diffusion through the solid domains are problematic [30,66]. However, when the pressure field is of direct interest, the Brinkman penalisation must be significantly higher to ensure an accurate evaluation [171].

For density-based methods and Ersatz-based level set methods on regular meshes, a smooth transition region, as ensured by e.g., density filtering, provides proper convergence of the design description with mesh refinement. It might not be an advantage to have a fully discrete 0–1 design field, since this will lead to staircase-like descriptions of the fluid–solid interface. For coarse meshes, this may lead to flow instabilities near the interface and thus a poor description of the boundary layer. For finer meshes, this is not as big of an issue.

4.3. Level Set-Based Approaches

The level set method is often praised for its accurate description of the geometric interface between solid and fluid. However, if this accurate description of the interface is not transferred to the simulation model, then nothing is gained. Therefore, in order to exploit the full potential of a level set design description, it becomes essential to use surface-capturing schemes, such as e.g., X-FEM [30,150], CutFEM [76], or adaptive body-fitted meshes [89,174]. This will allow for increased accuracy of the boundary layer, which becomes increasingly important when moving to more complex fluid problems, such as turbulent flow discussed in Section 4.7. Therefore, it is recommended that future work using the level set design representation should focus on applying surface-capturing schemes with local refinement of the boundary layer regions.

4.4. Steady-State Laminar Incompressible Flow

A steady-state incompressible laminar flow model is used for the vast majority of the work on topology optimisation of fluid-based problems, with 85% of all papers and 74% of fluid flow only papers. With 61 papers treating steady-state incompressible fluid flow only, it is proposed that the community not spend more time on this, especially for minimum dissipated energy and pressure drop. A large range of methods have been applied to these energy-based functionals, with only a minority of papers treating more complex objective functionals such as flow distribution and uniformity [33,50], diodicity [43,53,64,68,70,91], minimum drag and maximum lift [29,59].

Future papers treating only steady-state incompressible fluid flow should either present novel objective functionals, constraints or applications. This should preferably be in the context of application to practical engineering applications, since the treatment of steady-state incompressible laminar flow is already rather mature.

4.5. Benchmarking

Future papers should build and improve upon the current literature, not reproduce it. During the development and testing phases of research, already published examples should absolutely be used as benchmarks. However, merely reproducing old examples using a new method does not represent a scientific contribution. Future papers proposing new methodologies should show clear improvements compared to the old, focusing on the extension of applicability rather than reproducing old examples. Therefore, if a new method is not or can not be shown to provide a clear improvement in one or more of the following, the work does not warrant publication:

- accuracy of the geometric representation
- precision of solution and/or optimality
- algorithmic and/or computational efficiency
- parameter robustness and algorithmic stability

If the above is not shown, then the work should not be submitted by the authors and should be rejected by reviewers. Works reinventing the wheel with a new methodology without showing *clear* advantages, only serves to clog up the cogs of scientific progress.

In extension of the above, when comparing methodologies in order to show a clear improvement, it is pertinent to use the exact problem setup of the previously published works. There is a tendency to

4.6. Time-Dependent Problems

Only 13 papers in total treat time-dependent problems [72–79,145,150,152,158,188], eight of which are for fluid flow only. Since most realistic flow applications exhibit some form of time-dependent behaviour through either time-dependent boundary conditions or flow instabilities, it is strongly recommended that more community effort is dedicated to expanding the research on applying topology optimisation to time-dependent fluid-based problems. Due to the iterative nature of topology optimisation often requiring hundreds or thousands of simulations, the computational cost of a single time-dependent simulation becomes a significant bottleneck. The topology optimisation of time-dependent problems is therefore seen as the next frontier, requiring research into high-performance computing, efficient numerical methods and time integration and storage reduction methods. Novel ways to treat transient optimisation problems, such as the work by Chen et al. [77], can also aid in this progress. The topology optimisation community should draw inspiration from other fields, such as computational science and mathematics, and collaborate with researchers from those fields.

4.7. Turbulent Flow

Most industrial flow applications are turbulent, rather than laminar. Turbulence is inherently time-dependent and this research area goes hand in hand with the above. Current works on topology optimisation of turbulent flow only amount to six papers [21,80–83,129] and they all consider a steady-state time-averaged approximation of turbulence, namely the Reynolds-Averaged Navier–Stokes (RANS) equations. This is a natural starting point and there is certainly still room for research to be done at this level of turbulence modelling for topology optimisation.

Capturing the turbulent boundary layers is a significant challenge in topology optimisation, since the solid–fluid interface is not known a priori and, thus, local boundary layer mesh refinement is not easily applied. This can potentially be a significant problem for density-based methods with a gradual transition from solid to fluid or with a staircase description of the boundary. Surface-capturing level set methods can potentially deliver significant benefits to this type of problems, as well as adaptive body-conforming meshes [63,174] or local mesh refinement [61]. However, despite the attractive properties of accurate boundary identification, to date, only density-based methods for turbulent flow have been presented.

As will be discussed in Section 4.11, the introduction of approximate models as a surrogate for full-blown turbulent models may also be a viable way to treat very complex flow problems in the context of topology optimisation.

4.8. Compressible Flow

To the knowledge of the authors, no works treating fully compressible flows have been published as to the date of submission of this review paper. There are a few papers treating slightly compressible fluids, but only as an approximation of fully incompressible fluids. For this type of problem, local conservation properties may well prove important when introducing a varying design representation and, thus, methods such as FVM or Discontinuous Galerkin (DG-)FEM might be necessary to ensure conservation of mass.

4.9. Fluid–Structure Interaction

The efforts within wet topology optimisation of FSI problems only cover 15 papers [161,163–174], and these all remain restricted to small deformations and steady-state. Thus, the solution of the problems in the deformed state is either negligible or of minor importance to the optimisation procedure, at least if the design objective is minimum compliance. There seems to be a large potential in extending the methodology to transient problems exhibiting large deformations, e.g., in a biomechanical context.

4.10. Three-Dimensional Problems

As for time-dependent and turbulent problems, three-dimensionality is present in most industrial applications. Therefore, it is important for the community to focus on large scale three-dimensional problems. While 31% of papers treat three-dimensional problems, many of these suffer from either: being very small in the third dimension [107,122]; having severely restricted design freedom in the third dimension [164,168,172,173]; and using very coarse discretisations [14,23,46,56,132,150]. Truly three-dimensional problems inherently carry a large computational cost with them, and this is discussed in a number of papers, where high performance parallel computing [15,25,33,55,76,82,126,129,151,153,160], graphics processing units [119,173] and adaptive meshes [61] have been proposed as solutions.

Future papers treating three-dimensional problems should include a discussion of the computational cost involved. Since the research community should be moving towards more complicated flow problems including transient and turbulent flows, the concern of computational cost becomes even more dominant. Thus, even though simple problems may be treated in future papers, the computational cost and limitations of the method must be discussed in the context of tackling large scale three-dimensional problems.

4.11. Simplified Models or Approximations

Since all of the above problem areas all carry a large computational cost, it is beneficial for the community to work on simplified models or approximations to the complex physics.

4.11.1. 2D Simplification of 3D

It is very common in the included papers to treat two-dimensional academic problems. However, some works directly approximate a three-dimensional plane problem, with a small thickness, using a two-dimensional simplification, either stated explicitly [43,64,67,68,106,112,116,134,140,142,143,149] or implicitly [50,67,87,102,107]. Out of these, only three papers [106,140,143] include the viscous resistance from the friction due to the out-of-plane viscous boundary layers. This is despite the fact that a simple expression is given in the original works on the subject [7,8]. If the out-of-plane dimension is large, e.g., [152,158], the friction will go to zero. However, for small thicknesses, the friction cannot be neglected [143].

For forced convection cooling of heat sinks, pseudo-3D models have been proposed [113,127,128,141,143] consisting of two layers in order to approximate the temperature of both the heat source and a cross-section of the heat sink. Furthermore, a cross-sectional model for forced convection has also been proposed for flow that is fully-developed in the out-of-plane direction [121].

Common to all of the above dimensional simplifications is that the design is assumed to be constant in the out-of-plane direction and physical fields are assumed to vary polynomially in the out-of-plane direction. However, as shown by Dilgen et al. [129], the error introduced by this assumption can be rather large and, therefore, designs must be validated.

4.11.2. Simplified Flow Models

A number of the included papers use a simplified flow model compared to the situation that they wish to model. One examples is to use Stokes flow instead of turbulent flow, e.g., [116]; however, this is severely limited in capturing the correct physics. One suggestion to approximate the thin boundary layers for turbulent flow is by instead using a Darcy flow model [123] with an artificial permeability in the fluid region. However, inertia is still not captured. Another recent example uses laminar flow to approximate turbulent flow in a multifidelity approximation framework [148]. For natural convection, a potential-like model is derived by reducing the Navier–Stokes equations by assuming the buoyancy term to be dominant [157,160].

Using simplified models to treat complex problems is one way to reduce the computational cost, but it is extremely important to validate the design performance for the final design using the real model. Neglecting terms and phenomena leads to lower accuracy, but combining the simplified and full models in a sequential optimisation approach can reduce the cost significantly, while still retaining the accuracy of the full model for some steps of the optimisation. In engineering practise, a low-fidelity model is often used in the initial stages to make fast design progress and then a high-fidelity model is used to refine the design at the end [148]. However, Pollini et al. [160] recently proposed a sequential optimisation approach using the full model initially to point the gradient-based optimiser in the correct direction and then refine the design features using an approximate model.

4.12. Numerical Verification

For all papers treating the topology optimisation of fluid-based problems, numerical verification must be performed for the final design using an independent solver with a body-fitted mesh, sufficient mesh resolution and a fully descriptive physical model. This is a bare minimum for all future papers.

It is especially important for density- or Ersatz-material based approaches, where the boundary is not necessarily captured accurately on regular meshes. It is also important if the mesh used for optimisation is relatively coarse or where a simplified or reduced model has been used to ensure fast computations.

4.13. Experimental Validation

In addition to numerical verification, it is strongly suggested that, if at all possible, experimental validation is carried out, due to the complex geometries and complex physics encountered after fluid-based topology optimisation. One thing is that a simulation tool shows that the optimised geometry performs better than a reference design but should preferably be validated experimentally.

Only 12 papers, or 7%, contain some form of experimental investigation of the topology-optimised designs. These cover fixed-geometry fluid diodes [43,68], small scale rotary pumps [62], forced convection heat sinks [116,128,141,146], passive coolers for light-emitting diode lamps [154,155], porous bioreactors [186], particle separators [102], and conformal cooling channels [142].

5. Conclusions

This review paper provides an overview of the development of topology optimisation for fluid flow and fluid-based problems. Since the seminal paper by Borrvall and Petersson [7] in 2003, 186 papers have been published treating a large variety of phenomena and component design, ranging from creeping flow in pipes and microfluidic mixers to turbulent flow and heat transfer, from steady to time-dependent flows, and from simple academic problems to real-life industrial examples. A wide range of topology optimisation methods have been applied to the field, with most being classified as density-based methods and significantly less using level set methods. This is surprising since the potential strength of level set methods, in combination with surface-capturing discretisation schemes, is to provide a better definition of the interface, which can be necessary for more complicated fluid problems. On the contrary, the limited ability to create a new topology favours the density-based methods.

Recommendations for future research directions are outlined based on the extensive literature review, the quantitative analysis, as well as the authors' personal experience and opinions. The community is encouraged to focus on moving the field to more complicated applications, such as transient, turbulent and compressible flows. Generally, previously published examples should serve only the purpose of benchmarks for verifying a new method or implementation. However, if the new method does not show clear improvements in accuracy and efficiency over the previously published works, it should not be published. It is suggested that published works in the future only present improvements and extensions to the previous .

Author Contributions: The two authors have contributed equally to this work. All authors have read and agreed to the published version of the manuscript.

Funding: The second author was partly funded by the InnoTop project through the Villum Foundation.

Conflicts of Interest: The authors declare no conflicts of interest.

References

1. Bendsøe, M.P.; Kikuchi, N. Generating optimal topologies in structural design using a homogenization method. *Comput. Methods Appl. Mech. Eng.* **1988**, *71*, 197–224. [CrossRef]
2. Sigmund, O.; Maute, K. Topology optimization approaches: A comparative review. *Struct. Multidiscip. Optim.* **2013**, *48*, 1031–1055. [CrossRef]
3. Deaton, J.D.; Grandhi, R.V. A survey of structural and multidisciplinary continuum topology optimization: post 2000. *Struct. Multidiscip. Optim.* **2014**, *49*, 1–38. [CrossRef]
4. Chen, X. Topology optimization of microfluidics—A review. *Microchem. J.* **2016**, *127*, 52–61. [CrossRef]
5. Dbouk, T. A review about the engineering design of optimal heat transfer systems using topology optimization. *Appl. Therm. Eng.* **2017**, *112*, 841–854. [CrossRef]
6. van Dijk, N.P.; Maute, K.; Langelaar, M.; van Keulen, F. Level-set methods for structural topology optimization: A review. *Struct. Multidiscip. Optim.* **2013**, *48*, 437–472. [CrossRef]
7. Borrvall, T.; Petersson, J. Topology optimization of fluids in Stokes flow. *Int. J. Numer. Methods Fluids* **2003**, *41*, 77–107. [CrossRef]
8. Gersborg-Hansen, A.; Sigmund, O.; Haber, R.B. Topology optimization of channel flow problems. *Struct. Multidiscip. Optim.* **2005**, *30*, 181–192. [CrossRef]
9. Evgrafov, A. The Limits of Porous Materials in the Topology Optimization of Stokes Flows. *Appl. Math. Optim.* **2005**, *52*, 263–277. [CrossRef]
10. Olesen, L.H.; Okkels, F.; Bruus, H. A high-level programming-language implementation of topology optimization applied to steady-state Navier–Stokes flow. *Int. J. Numer. Methods Eng.* **2006**, *65*, 975–1001. [CrossRef]
11. Guest, J.K.; Prévost, J.H. Topology optimization of creeping fluid flows using a Darcy–Stokes finite element. *Int. J. Numer. Methods Eng.* **2006**, *66*, 461–484. [CrossRef]
12. Evgrafov, A. Topology optimization of slightly compressible fluids. *ZAMM* **2006**, *86*, 46–62. [CrossRef]
13. Wiker, N.; Klarbring, A.; Borrvall, T. Topology optimization of regions of Darcy and Stokes flow. *Int. J. Numer. Methods Eng.* **2007**, *69*, 1374–1404. [CrossRef]
14. Pingen, G.; Evgrafov, A.; Maute, K. Topology optimization of flow domains using the lattice Boltzmann method. *Struct. Multidiscip. Optim.* **2007**, *34*, 507–524. [CrossRef]
15. Aage, N.; Poulsen, T.H.; Gersborg-Hansen, A.; Sigmund, O. Topology optimization of large scale Stokes flow problems. *Struct. Multidiscip. Optim.* **2007**, *35*, 175–180. [CrossRef]
16. Bruns, T.E. Topology optimization by penalty (TOP) method. *Comput. Methods Appl. Mech. Eng.* **2007**, *196*, 4430–4443. [CrossRef]
17. Duan, X.B.; Ma, Y.C.; Zhang, R. Shape-topology optimization for Navier–Stokes problem using variational level set method. *J. Comput. Appl. Math.* **2008**, *222*, 487–499. [CrossRef]
18. Duan, X.B.; Ma, Y.C.; Zhang, R. Shape-topology optimization of Stokes flow via variational level set method. *Appl. Math. Comput.* **2008**, *202*, 200–209. doi:10.1016/j.amc.2008.02.014. [CrossRef]
19. Duan, X.; Ma, Y.; Zhang, R. Optimal shape control of fluid flow using variational level set method. *Phys. Lett. A* **2008**, *372*, 1374–1379. doi:10.1016/j.physleta.2007.09.070. [CrossRef]
20. Evgrafov, A.; Pingen, G.; Maute, K. Topology optimization of fluid domains: kinetic theory approach. *ZAMM* **2008**, *88*, 129–141. [CrossRef]
21. Othmer, C. A continuous adjoint formulation for the computation of topological and surface sensitivities of ducted flows. *Int. J. Numer. Methods Fluids* **2008**, *58*, 861–877. [CrossRef]

22. Pingen, G.; Evgrafov, A.; Maute, K. A parallel Schur complement solver for the solution of the adjoint steady-state lattice Boltzmann equations: application to design optimisation. *Int. J. Comput. Fluid Dyn.* **2008**, *22*, 457–464. [CrossRef]
23. Zhou, S.; Li, Q. A variational level set method for the topology optimization of steady-state Navier–Stokes flow. *J. Comput. Phys.* **2008**, *227*, 10178–10195. [CrossRef]
24. Pingen, G.; Waidmann, M.; Evgrafov, A.; Maute, K. A parametric level-set approach for topology optimization of flow domains. *Struct. Multidiscip. Optim.* **2010**, *41*, 117–131. [CrossRef]
25. Challis, V.J.; Guest, J.K. Level set topology optimization of fluids in Stokes flow. *Int. J. Numer. Methods Eng.* **2009**, *79*, 1284–1308. [CrossRef]
26. Kreissl, S.; Pingen, G.; Maute, K. An explicit level set approach for generalized shape optimization of fluids with the lattice Boltzmann method. *Int. J. Numer. Methods Fluids* **2011**, *65*, 496–519. [CrossRef]
27. Liu, Z.; Gao, Q.; Zhang, P.; Xuan, M.; Wu, Y. Topology optimization of fluid channels with flow rate equality constraints. *Struct. Multidiscip. Optim.* **2011**, *44*, 31–37. [CrossRef]
28. Okkels, F.; Dufva, M.; Bruus, H. Optimal homogenization of perfusion flows in microfluidic bio-reactors: A numerical study. *PLoS ONE* **2011**, *6*, e14574. [CrossRef]
29. Kondoh, T.; Matsumori, T.; Kawamoto, A. Drag minimization and lift maximization in laminar flows via topology optimization employing simple objective function expressions based on body force integration. *Struct. Multidiscip. Optim.* **2012**, *45*, 693–701. [CrossRef]
30. Kreissl, S.; Maute, K. Levelset based fluid topology optimization using the extended finite element method. *Struct. Multidiscip. Optim.* **2012**, *46*, 311–326. [CrossRef]
31. Deng, Y.; Liu, Z.; Wu, Y. Topology optimization of steady and unsteady incompressible Navier–Stokes flows driven by body forces. *Struct. Multidiscip. Optim.* **2013**, *47*, 555–570. [CrossRef]
32. Deng, Y.; Liu, Z.; Wu, J.; Wu, Y. Topology optimization of steady Navier–Stokes flow with body force. *Comput. Methods Appl. Mech. Eng.* **2013**, *255*, 306–321. [CrossRef]
33. Aage, N.; Lazarov, B.S. Parallel framework for topology optimization using the method of moving asymptotes. *Struct. Multidiscip. Optim.* **2013**, *47*, 493–505. [CrossRef]
34. Evgrafov, A. State space Newton's method for topology optimization. *Comput. Methods Appl. Mech. Eng.* **2014**, *278*, 272–290. [CrossRef]
35. Romero, J.S.; Silva, E.C.N. A topology optimization approach applied to laminar flow machine rotor design. *Comput. Methods Appl. Mech. Eng.* **2014**, *279*, 268–300. [CrossRef]
36. Yaji, K.; Yamada, T.; Yoshino, M.; Matsumoto, T.; Izui, K.; Nishiwaki, S. Topology optimization using the lattice Boltzmann method incorporating level set boundary expressions. *J. Comput. Phys.* **2014**, *274*, 158–181. [CrossRef]
37. Liu, Z.; Liu, G.; Geier, M.; Krafczyk, M.; Chen, T. Discrete adjoint sensitivity analysis for fluid flow topology optimization based on the generalized lattice Boltzmann method. *Comput. Math. Appl.* **2014**, *68*, 1374–1392. [CrossRef]
38. Yonekura, K.; Kanno, Y. A flow topology optimization method for steady state flow using transient information of flow field solved by lattice Boltzmann method. *Struct. Multidiscip. Optim.* **2015**, *51*, 159–172. [CrossRef]
39. Liu, X.; Zhang, B.; Sun, J. An improved implicit re-initialization method for the level set function applied to shape and topology optimization of fluid. *J. Comput. Appl. Math.* **2015**, *281*, 207–229. [CrossRef]
40. Duan, X.B.; Li, F.F.; Qin, X.Q. Adaptive mesh method for topology optimization of fluid flow. *Appl. Math. Lett.* **2015**, *44*, 40–44. [CrossRef]
41. Evgrafov, A. On Chebyshev's method for topology optimization of Stokes flows. *Struct. Multidiscip. Optim.* **2015**, *51*, 801–811. [CrossRef]
42. Garcke, H.; Hecht, C. Shape and Topology Optimization in Stokes Flow with a Phase Field Approach. *Appl. Math. Optim.* **2016**, *73*, 23–70. [CrossRef]
43. Lin, S.; Zhao, L.; Guest, J.K.; Weihs, T.P.; Liu, Z. Topology Optimization of Fixed-Geometry Fluid Diodes. *J. Mech. Des.* **2015**, *137*, 081402. [CrossRef]
44. Garcke, H.; Hecht, C.; Hinze, M.; Kahle, C. Numerical Approximation of Phase Field Based Shape and Topology Optimization for Fluids. *SIAM J. Sci. Comput.* **2015**, *37*, A1846–A1871. [CrossRef]
45. Duan, X.; Qin, X.; Li, F. Topology optimization of Stokes flow using an implicit coupled level set method. *Appl. Math. Model.* **2016**, *40*, 5431–5441. [CrossRef]

46. N. Sá, L.F.; R. Amigo, R.C.; Novotny, A.A.; N. Silva, E.C. Topological derivatives applied to fluid flow channel design optimization problems. *Struct. Multidiscip. Optim.* **2016**, *54*, 249–264. [CrossRef]
47. Yoshimura, M.; Shimoyama, K.; Misaka, T.; Obayashi, S. Topology optimization of fluid problems using genetic algorithm assisted by the Kriging model. *Int. J. Numer. Methods Eng.* **2017**, *109*, 514–532. [CrossRef]
48. Pereira, A.; Talischi, C.; Paulino, G.H.; M. Menezes, I.F.; Carvalho, M.S. Fluid flow topology optimization in PolyTop: stability and computational implementation. *Struct. Multidiscip. Optim.* **2016**, *54*, 1345–1364. [CrossRef]
49. Duan, X.; Li, F.; Qin, X. Topology optimization of incompressible Navier–Stokes problem by level set based adaptive mesh method. *Comput. Math. Appl.* **2016**, *72*, 1131–1141. [CrossRef]
50. Kubo, S.; Yaji, K.; Yamada, T.; Izui, K.; Nishiwaki, S. A level set-based topology optimization method for optimal manifold designs with flow uniformity in plate-type microchannel reactors. *Struct. Multidiscip. Optim.* **2017**, *55*, 1311–1327. [CrossRef]
51. Jang, G.W.; Lee, J.W. Topology optimization of internal partitions in a flow-reversing chamber muffler for noise reduction. *Struct. Multidiscip. Optim.* **2016**, *55*, 2181–2196. [CrossRef]
52. Koch, J.R.L.; Papoutsis-Kiachagias, E.M.; Giannakoglou, K.C. Transition from adjoint level set topology to shape optimization for 2D fluid mechanics. *Comput. Fluids* **2017**, *150*, 123–138. [CrossRef]
53. Sato, Y.; Yaji, K.; Izui, K.; Yamada, T.; Nishiwaki, S. Topology optimization of a no-moving-part valve incorporating Pareto frontier exploration. *Struct. Multidiscip. Optim.* **2017**, *56*, 839–851. [CrossRef]
54. Sá, N.L.F.; Novotny, A.A.; Romero, J.S.; N. Silva, E.C. Design optimization of laminar flow machine rotors based on the topological derivative concept. *Struct. Multidiscip. Optim.* **2017**, *56*, 1013–1026. [CrossRef]
55. Yonekura, K.; Kanno, Y. Topology optimization method for interior flow based on transient information of the lattice Boltzmann method with a level-set function. *Jpn. J. Ind. Appl. Math.* **2017**, *34*, 611–632. [CrossRef]
56. Dai, X.; Zhang, C.; Zhang, Y.; Gulliksson, M. Topology optimization of steady Navier–Stokes flow via a piecewise constant level set method. *Struct. Multidiscip. Optim.* **2018**, *57*, 2193–2203. [CrossRef]
57. Shen, C.; Hou, L.; Zhang, E.; Lin, J. Topology optimization of three-phase interpolation models in Darcy–Stokes flow. *Struct. Multidiscip. Optim.* **2018**, *57*, 1663–1677. [CrossRef]
58. Deng, Y.; Liu, Z.; Wu, Y. Topology Optimization of Capillary, Two-Phase Flow Problems. *Commun. Comput. Phys.* **2017**, *22*, 1413–1438. [CrossRef]
59. Garcke, H.; Hinze, M.; Kahle, C.; Lam, K.F. A phase field approach to shape optimization in Navier–Stokes flow with integral state constraints. *Adv. Comput. Math.* **2018**, *44*, 1345–1383. [CrossRef]
60. Alonso, D.H.; de Sá, L.F.N.; Saenz, J.S.R.; Silva, E.C.N. Topology optimization applied to the design of 2D swirl flow devices. *Struct. Multidiscip. Optim.* **2018**, *58*, 2341–2364. [CrossRef]
61. Jensen, K.E. Topology optimization of Stokes flow on dynamic meshes using simple optimizers. *Comput. Fluids* **2018**, *174*, 66–77. [CrossRef]
62. Sá, L.F.N.; Romero, J.S.; Horikawa, O.; Silva, E.C.N. Topology optimization applied to the development of small scale pump. *Struct. Multidiscip. Optim.* **2018**, *57*, 2045–2059. [CrossRef]
63. Zhou, M.; Lian, H.; Sigmund, O.; Aage, N. Shape morphing and topology optimization of fluid channels by explicit boundary tracking. *Int. J. Numer. Methods Fluids* **2018**, *88*, 296–313. [CrossRef]
64. Shin, S.; Jeong, J.H.; Lim, D.K.; Kim, E.S. Design of SFR fluidic diode axial port using topology optimization. *Nucl. Eng. Des.* **2018**, *338*, 63–73. [CrossRef]
65. Yonekura, K.; Kanno, Y. A Heuristic Method Using Hessian Matrix for Fast Flow Topology Optimization. *J. Optim. Theory Appl.* **2019**, *180*, 671–681. [CrossRef]
66. Behrou, R.; Ranjan, R.; Guest, J.K. Adaptive topology optimization for incompressible laminar flow problems with mass flow constraints. *Comput. Methods Appl. Mech. Eng.* **2019**, *346*, 612–641. [CrossRef]
67. Alonso, D.H.; de Sá, L.F.N.; Saenz, J.S.R.; Silva, E.C.N. Topology optimization based on a two-dimensional swirl flow model of Tesla-type pump devices. *Comput. Math. Appl.* **2019**, *77*, 2499–2533. [CrossRef]
68. Lim, D.K.; Song, M.S.; Chae, H.; Kim, E.S. Topology optimization on vortex-type passive fluidic diode for advanced nuclear reactors. *Nucl. Eng. Technol.* **2019**, *51*, 1279–1288. [CrossRef]
69. Sato, A.; Yamada, T.; Izui, K.; Nishiwaki, S.; Takata, S. A topology optimization method in rarefied gas flow problems using the Boltzmann equation. *J. Comput. Phys.* **2019**, *395*, 60–84. [CrossRef]
70. Gaymann, A.; Montomoli, F.; Pietropaoli, M. Fluid topology optimization: Bio-inspired valves for aircraft engines. *Int. J. Heat Fluid Flow* **2019**, *79*, 108455. [CrossRef]

71. Gaymann, A.; Montomoli, F. Deep Neural Network and Monte Carlo Tree Search applied to fluid–structure Topology Optimization. *Sci. Rep.* **2019**, *9*, 15916. [CrossRef] [PubMed]
72. Kreissl, S.; Pingen, G.; Maute, K. Topology optimization for unsteady flow. *Int. J. Numer. Methods Eng.* **2011**, *87*, 1229–1253. [CrossRef]
73. Deng, Y.; Liu, Z.; Zhang, P.; Liu, Y.; Wu, Y. Topology optimization of unsteady incompressible Navier–Stokes flows. *J. Comput. Phys.* **2011**, *230*, 6688–6708. [CrossRef]
74. Abdelwahed, M.; Hassine, M. Topology Optimization of Time Dependent Viscous Incompressible Flows. *Abstr. Appl. Anal.* **2014**, *2014*, 923016. [CrossRef]
75. Nørgaard, S.; Sigmund, O.; Lazarov, B. Topology optimization of unsteady flow problems using the lattice Boltzmann method. *J. Comput. Phys.* **2016**, *307*, 291–307. [CrossRef]
76. Villanueva, C.H.; Maute, K. CutFEM topology optimization of 3D laminar incompressible flow problems. *Comput. Methods Appl. Mech. Eng.* **2017**, *320*, 444–473. [CrossRef]
77. Chen, C.; Yaji, K.; Yamada, T.; Izui, K.; Nishiwaki, S. Local-in-time adjoint-based topology optimization of unsteady fluid flows using the lattice Boltzmann method. *Mech. Eng. J.* **2017**, *4*, 17-00120. [CrossRef]
78. Nørgaard, S.A.; Sagebaum, M.; Gauger, N.R.; Lazarov, B.S. Applications of automatic differentiation in topology optimization. *Struct. Multidiscip. Optim.* **2017**, *56*, 1135–1146. [CrossRef]
79. Sasaki, Y.; Sato, Y.; Yamada, T.; Izui, K.; Nishiwaki, S. Topology optimization for fluid flows using the MPS method incorporating the level set method. *Comput. Fluids* **2019**, *188*, 86–101. [CrossRef]
80. Kontoleontos, E.A.; Papoutsis-Kiachagias, E.M.; Zymaris, A.S.; Papadimitriou, D.I.; Giannakoglou, K.C. Adjoint-based constrained topology optimization for viscous flows, including heat transfer. *Eng. Optim.* **2013**, *45*, 941–961. [CrossRef]
81. Yoon, G.H. Topology optimization for turbulent flow with Spalart–Allmaras model. *Comput. Methods Appl. Mech. Eng.* **2016**, *303*, 288–311. [CrossRef]
82. Dilgen, S.B.; Dilgen, C.B.; Fuhrman, D.R.; Sigmund, O.; Lazarov, B.S. Density based topology optimization of turbulent flow heat transfer systems. *Struct. Multidiscip. Optim.* **2018**, *57*, 1905–1918. [CrossRef]
83. Yoon, G.H. Topology optimization method with finite elements based on the k-ϵ turbulence model. *Comput. Methods Appl. Mech. Eng.* **2020**, *361*, 112784. [CrossRef]
84. Pingen, G.; Maute, K. Optimal design for non-Newtonian flows using a topology optimization approach. *Comput. Math. Appl.* **2010**, *59*, 2340–2350. [CrossRef]
85. Ejlebjerg Jensen, K.; Szabo, P.; Okkels, F. Topology optimization of viscoelastic rectifiers. *Appl. Phys. Lett.* **2012**, *100*, 234102. [CrossRef]
86. Jensen, K.E.; Szabo, P.; Okkels, F. Optimization of bistable viscoelastic systems. *Struct. Multidiscip. Optim.* **2014**, *49*, 733–742. [CrossRef]
87. Hyun, J.; Wang, S.; Yang, S. Topology optimization of the shear thinning non-Newtonian fluidic systems for minimizing wall shear stress. *Comput. Math. Appl.* **2014**, *67*, 1154–1170. [CrossRef]
88. Zhang, B.; Liu, X. Topology optimization study of arterial bypass configurations using the level set method. *Struct. Multidiscip. Optim.* **2015**, *51*, 773–798. [CrossRef]
89. Zhang, B.; Liu, X.; Sun, J. Topology optimization design of non-Newtonian roller-type viscous micropumps. *Struct. Multidiscip. Optim.* **2016**, *53*, 409–424. [CrossRef]
90. Romero, J.S.; N. Silva, E.C. Non-newtonian laminar flow machine rotor design by using topology optimization. *Struct. Multidiscip. Optim.* **2017**, *55*, 1711–1732. [CrossRef]
91. Dong, X.; Liu, X. Bi-objective topology optimization of asymmetrical fixed-geometry microvalve for non-Newtonian flow. *Microsyst. Technol.* **2019**, *25*, 2471–2479. [CrossRef]
92. Okkels, F.; Bruus, H. Scaling behavior of optimally structured catalytic microfluidic reactors. *Phys. Rev. E* **2007**, *75*, 016301. [CrossRef] [PubMed]
93. Andreasen, C.S.; Gersborg, A.R.; Sigmund, O. Topology optimization of microfluidic mixers. *Int. J. Numer. Methods Fluids* **2009**, *61*, 498–513. [CrossRef]
94. Gregersen, M.M.; Okkels, F.; Bazant, M.Z.; Bruus, H. Topology and shape optimization of induced-charge electro-osmotic micropumps. *New J. Phys.* **2009**, *11*, 075019. [CrossRef]
95. Schäpper, D.; Lencastre Fernandes, R.; Lantz, A.E.; Okkels, F.; Bruus, H.; Gernaey, K.V. Topology optimized microbioreactors. *Biotechnol. Bioeng.* **2011**, *108*, 786–796. [CrossRef] [PubMed]
96. Kim, C.; Sun, H. Topology optimization of gas flow channel routes in an automotive fuel cell. *Int. J. Automot. Technol.* **2012**, *13*, 783–789. [CrossRef]

97. Makhija, D.; Pingen, G.; Yang, R.; Maute, K. Topology optimization of multi-component flows using a multi-relaxation time lattice Boltzmann method. *Comput. Fluids* **2012**, *67*, 104–114. [CrossRef]
98. Deng, Y.; Liu, Z.; Zhang, P.; Liu, Y.; Gao, Q.; Wu, Y. A flexible layout design method for passive micromixers. *Biomed. Microdevices* **2012**, *14*, 929–945. [CrossRef]
99. Makhija, D.; Maute, K. Level set topology optimization of scalar transport problems. *Struct. Multidiscip. Optim.* **2014**, *51*, 267–285. [CrossRef]
100. Oh, S.; Wang, S.; Park, M.; Kim, J.H. Novel spacer design using topology optimization in a reverse osmosis channel. *J. Fluids Eng. Trans. ASME* **2014**, *136*, 021201. [CrossRef]
101. Chen, X.; Li, T. A novel design for passive misscromixers based on topology optimization method. *Biomed. Microdevices* **2016**, *18*, 57. [CrossRef]
102. Hyun, J.c.; Hyun, J.; Wang, S.; Yang, S. Improved pillar shape for deterministic lateral displacement separation method to maintain separation efficiency over a long period of time. *Sep. Purif. Technol.* **2017**, *172*, 258–267. [CrossRef]
103. Andreasen, C.S. Topology optimization of inertia driven dosing units. *Struct. Multidiscip. Optim.* **2016**, *55*, 1301–1309. [CrossRef]
104. Yaji, K.; Yamasaki, S.; Tsushima, S.; Suzuki, T.; Fujita, K. Topology optimization for the design of flow fields in a redox flow battery. *Struct. Multidiscip. Optim.* **2018**, *57*, 535–546. [CrossRef]
105. Guo, Y.; Xu, Y.; Deng, Y.; Liu, Z. Topology Optimization of Passive Micromixers Based on Lagrangian Mapping Method. *Micromachines* **2018**, *9*, 137. [CrossRef] [PubMed]
106. Behrou, R.; Pizzolato, A.; Forner-Cuenca, A. Topology optimization as a powerful tool to design advanced PEMFCs flow fields. *Int. J. Heat Mass Transf.* **2019**, *135*, 72–92. [CrossRef]
107. Chen, C.H.; Yaji, K.; Yamasaki, S.; Tsushima, S.; Fujita, K. Computational design of flow fields for vanadium redox flow batteries via topology optimization. *J. Energy Storage* **2019**, *26*, 100990. [CrossRef]
108. Dugast, F.; Favenne, Y.; Josset, C. Reactive fluid flow topology optimization with the multi-relaxation time lattice Boltzmann method and a level-set function. *J. Comput. Phys.* **2020**, 109252. [CrossRef]
109. Alexandersen, J.; Aage, N.; Andreasen, C.S.; Sigmund, O. Topology optimisation for natural convection problems. *Int. J. Numer. Methods Fluids* **2014**, *76*, 699–721. [CrossRef]
110. Dede, E.M. Multiphysics Topology Optimization of Heat Transfer and Fluid Flow Systems. In Proceedings of the COMSOL Conference 2009, Boston, MA, USA, 8–10 October 2009.
111. Yoon, G.H. Topological design of heat dissipating structure with forced convective heat transfer. *J. Mech. Sci. Technol.* **2010**, *24*, 1225–1233. [CrossRef]
112. Dede, E.M. Optimization and design of a multipass branching microchannel heat sink for electronics cooling. *J. Electron. Packag. Trans. ASME* **2012**, *134*, 041001. [CrossRef]
113. McConnell, C.; Pingen, G. Multi-Layer, Pseudo 3D Thermal Topology Optimization of Heat Sinks. In Proceedings of the ASME 2012 International Mechanical Engineering Congress and Exposition, Houston, TX, USA, 9–15 November 2012; pp. 2381–2392.
114. Matsumori, T.; Kondoh, T.; Kawamoto, A.; Nomura, T. Topology optimization for fluid–thermal interaction problems under constant input power. *Struct. Multidiscip. Optim.* **2013**, *47*, 571–581. [CrossRef]
115. Marck, G.; Nemer, M.; Harion, J.L. Topology Optimization of Heat and Mass Transfer Problems: Laminar Flow. *Numer. Heat Transf. Part B Fundam.* **2013**, *63*, 508–539. [CrossRef]
116. Koga, A.A.; Lopes, E.C.C.; Villa Nova, H.F.; Lima, C.R.d.; Silva, E.C.N. Development of heat sink device by using topology optimization. *Int. J. Heat Mass Transf.* **2013**, *64*, 759–772. [CrossRef]
117. Yaji, K.; Yamada, T.; Kubo, S.; Izui, K.; Nishiwaki, S. A topology optimization method for a coupled thermal–fluid problem using level set boundary expressions. *Int. J. Heat Mass Transf.* **2015**, *81*, 878–888. [CrossRef]
118. Yaji, K.; Yamada, T.; Yoshino, M.; Matsumoto, T.; Izui, K.; Nishiwaki, S. Topology optimization in thermal-fluid flow using the lattice Boltzmann method. *J. Comput. Phys.* **2016**, *307*, 355–377. [CrossRef]
119. Łaniewski Wołłk, L.; Rokicki, J. Adjoint Lattice Boltzmann for topology optimization on multi-GPU architecture. *Comput. Math. Appl.* **2016**, *71*, 833–848. [CrossRef]
120. Qian, X.; Dede, E.M. Topology optimization of a coupled thermal-fluid system under a tangential thermal gradient constraint. *Struct. Multidiscip. Optim.* **2016**, *54*, 531–551. [CrossRef]
121. Haertel, J.H.K.; Nellis, G.F. A fully developed flow thermofluid model for topology optimization of 3D-printed air-cooled heat exchangers. *Appl. Therm. Eng.* **2017**, *119*, 10–24. [CrossRef]

122. Pietropaoli, M.; Ahlfeld, R.; Montomoli, F.; Ciani, A.; D'Ercole, M. Design for Additive Manufacturing: Internal Channel Optimization. *J. Eng. Gas Turbines Power* **2017**, *139*, 102101. [CrossRef]
123. Zhao, X.; Zhou, M.; Sigmund, O.; Andreasen, C.S. A 'poor man's approach" to topology optimization of cooling channels based on a Darcy flow model. *Int. J. Heat Mass Transf.* **2018**, *116*, 1108–1123. [CrossRef]
124. Qian, S.; Wang, W.; Ge, C.; Lou, S.; Miao, E.; Tang, B. Topology optimization of fluid flow channel in cold plate for active phased array antenna. *Struct. Multidiscip. Optim.* **2018**, *57*, 2223–2232. [CrossRef]
125. Sato, Y.; Yaji, K.; Izui, K.; Yamada, T.; Nishiwaki, S. An Optimum Design Method for a Thermal-Fluid Device Incorporating Multiobjective Topology Optimization With an Adaptive Weighting Scheme. *J. Mech. Des.* **2018**, *140*, 31402. [CrossRef]
126. Yaji, K.; Ogino, M.; Chen, C.; Fujita, K. Large-scale topology optimization incorporating local-in-time adjoint-based method for unsteady thermal-fluid problem. *Struct. Multidiscip. Optim.* **2018**, *58*, 817–822. [CrossRef]
127. Haertel, J.H.K.; Engelbrecht, K.; Lazarov, B.S.; Sigmund, O. Topology optimization of a pseudo 3D thermofluid heat sink model. *Int. J. Heat Mass Transf.* **2018**, *121*, 1073–1088. [CrossRef]
128. Zeng, S.; Kanargi, B.; Lee, P.S. Experimental and numerical investigation of a mini channel forced air heat sink designed by topology optimization. *Int. J. Heat Mass Transf.* **2018**, *121*, 663–679. [CrossRef]
129. Dilgen, C.B.; Dilgen, S.B.; Fuhrman, D.R.; Sigmund, O.; Lazarov, B.S. Topology optimization of turbulent flows. *Comput. Methods Appl. Mech. Eng.* **2018**, *331*, 363–393. [CrossRef]
130. Ramalingom, D.; Cocquet, P.H.; Bastide, A. A new interpolation technique to deal with fluid-porous media interfaces for topology optimization of heat transfer. *Comput. Fluids* **2018**, *168*, 144–158. [CrossRef]
131. Dugast, F.; Favennec, Y.; Josset, C.; Fan, Y.; Luo, L. Topology optimization of thermal fluid flows with an adjoint Lattice Boltzmann Method. *J. Comput. Phys.* **2018**, *365*, 376–404. [CrossRef]
132. Santhanakrishnan, M.S.; Tilford, T.; Bailey, C. Performance assessment of density and level-set topology optimisation methods for three-dimensional heat sink design. *J. Algorithms Comput. Technol.* **2018**, *12*, 273–287. [CrossRef]
133. Sun, C.; Lewpiriyawong, N.; Khoo, K.L.; Zeng, S.; Lee, P.S. Thermal enhancement of fin and tube heat exchanger with guiding channels and topology optimisation. *Int. J. Heat Mass Transf.* **2018**, *127*, 1001–1013. [CrossRef]
134. Lv, Y.; Liu, S. Topology optimization and heat dissipation performance analysis of a micro-channel heat sink. *Meccanica* **2018**, *53*, 3693–3708. [CrossRef]
135. Pietropaoli, M.; Montomoli, F.; Gaymann, A. Three-dimensional fluid topology optimization for heat transfer. *Struct. Multidiscip. Optim.* **2019**, *59*, 801–812. [CrossRef]
136. Makhija, D.S.; Beran, P.S. Concurrent shape and topology optimization for steady conjugate heat transfer. *Struct. Multidiscip. Optim.* **2019**, *59*, 919–940. [CrossRef]
137. Subramaniam, V.; Dbouk, T.; Harion, J.L. Topology optimization of conjugate heat transfer systems: A competition between heat transfer enhancement and pressure drop reduction. *Int. J. Heat Fluid Flow* **2019**, *75*, 165–184. [CrossRef]
138. Yu, M.; Ruan, S.; Wang, X.; Li, Z.; Shen, C. Topology optimization of thermal–fluid problem using the MMC-based approach. *Struct. Multidiscip. Optim.* **2019**, *60*, 151–165. [CrossRef]
139. Zhang, B.; Gao, L. Topology optimization of convective heat transfer problems for non-Newtonian fluids. *Struct. Multidiscip. Optim.* **2019**, *60*, 1821–1840. [CrossRef]
140. Kobayashi, H.; Yaji, K.; Yamasaki, S.; Fujita, K. Freeform winglet design of fin-and-tube heat exchangers guided by topology optimization. *Appl. Therm. Eng.* **2019**, *161*, 114020. [CrossRef]
141. Zeng, S.; Lee, P.S. Topology optimization of liquid-cooled microchannel heat sinks: An experimental and numerical study. *Int. J. Heat Mass Transf.* **2019**, *142*, 118401. [CrossRef]
142. Jahan, S.; Wu, T.; Shin, Y.; Tovar, A.; El-Mounayri, H. Thermo-fluid Topology Optimization and Experimental Study of Conformal Cooling Channels for 3D Printed Plastic Injection Molds. *Procedia Manuf.* **2019**, *34*, 631–639. [CrossRef]
143. Yan, S.; Wang, F.; Hong, J.; Sigmund, O. Topology optimization of microchannel heat sinks using a two-layer model. *Int. J. Heat Mass Transf.* **2019**, *143*, 118462. [CrossRef]
144. Tawk, R.; Ghannam, B.; Nemer, M. Topology optimization of heat and mass transfer problems in two fluids-one solid domains. *Numer. Heat Transf. Part B Fundam.* **2019**, *76*, 130–151. [CrossRef]

145. Lundgaard, C.; Engelbrecht, K.; Sigmund, O. A density-based topology optimization methodology for thermal energy storage systems. *Struct. Multidiscip. Optim.* **2019**, *60*, 2189–2204. [CrossRef]
146. Li, H.; Ding, X.; Meng, F.; Jing, D.; Xiong, M. Optimal design and thermal modelling for liquid-cooled heat sink based on multi-objective topology optimization: An experimental and numerical study. *Int. J. Heat Mass Transf.* **2019**, *144*, 118638. [CrossRef]
147. Dong, X.; Liu, X. Multi-objective optimal design of microchannel cooling heat sink using topology optimization method. *Numeri. Heat Transf. Part A Appl.* **2020**, *77*, 90–104. [CrossRef]
148. Yaji, K.; Yamasaki, S.; Fujita, K. Multifidelity design guided by topology optimization. *Struct. Multidiscip. Optim.* **2019**, *61*, 1071–1085. [CrossRef]
149. Hu, D.; Zhang, Z.; Li, Q. Numerical study on flow and heat transfer characteristics of microchannel designed using topological optimizations method. *Sci. China Technol. Sci.* **2020**, *63*, 105–115. [CrossRef]
150. Coffin, P.; Maute, K. A level-set method for steady-state and transient natural convection problems. *Struct. Multidiscip. Optim.* **2016**, *53*, 1047–1067. [CrossRef]
151. Alexandersen, J.; Sigmund, O.; Aage, N. Large scale three-dimensional topology optimisation of heat sinks cooled by natural convection. *Int. J. Heat Mass Transf.* **2016**, *100*, 876–891. [CrossRef]
152. Pizzolato, A.; Sharma, A.; Maute, K.; Sciacovelli, A.; Verda, V. Design of effective fins for fast PCM melting and solidification in shell-and-tube latent heat thermal energy storage through topology optimization. *Appl. Energy* **2017**, *208*, 210–227. [CrossRef]
153. Alexandersen, J.; Sigmund, O.; Meyer, K.E.; Lazarov, B.S. Design of passive coolers for light-emitting diode lamps using topology optimisation. *Int. J. Heat Mass Transf.* **2018**, *122*, 138–149. [CrossRef]
154. Lazarov, B.S.; Sigmund, O.; Meyer, K.E.; Alexandersen, J. Experimental validation of additively manufactured optimized shapes for passive cooling. *Appl. Energy* **2018**, *226*, 330–339. [CrossRef]
155. Lei, T.; Alexandersen, J.; Lazarov, B.S.; Wang, F.; Haertel, J.H.K.; De Angelis, S.; Sanna, S.; Sigmund, O.; Engelbrecht, K. Investment casting and experimental testing of heat sinks designed by topology optimization. *Int. J. Heat Mass Transf.* **2018**, *127*, 396–412. [CrossRef]
156. Saglietti, C.; Schlatter, P.; Wadbro, E.; Berggren, M.; Henningson, D.S. Topology optimization of heat sinks in a square differentially heated cavity. *Int. J. Heat Fluid Flow* **2018**, *74*, 36–52. [CrossRef]
157. Asmussen, J.; Alexandersen, J.; Sigmund, O.; Andreasen, C.S. A "poor man's" approach to topology optimization of natural convection problems. *Struct. Multidiscip. Optim.* **2019**, *59*, 1105–1124. [CrossRef]
158. Pizzolato, A.; Sharma, A.; Ge, R.; Maute, K.; Verda, V.; Sciacovelli, A. Maximization of performance in multi-tube latent heat storage–Optimization of fins topology, effect of materials selection and flow arrangements. *Energy* **2019**, in press.
159. Ramalingom, D.; Cocquet, P.H.; Maleck, R.; Bastide, A. A multi-objective optimization problem in mixed and natural convection for a vertical channel asymmetrically heated. *Struct. Multidiscip. Optim.* **2019**, *60*, 2001–2020. [CrossRef]
160. Pollini, N.; Sigmund, O.; Andreasen, C.S.; Alexandersen, J. A "poor man's" approach for high-resolution three-dimensional topology design for natural convection problems. *Adv. Eng. Softw.* **2020**, *140*, 102736. doi:10.1016/j.advengsoft.2019.102736. [CrossRef]
161. Yoon, G.H. Topology optimization for stationary fluid–structure interaction problems using a new monolithic formulation. *Int. J. Numer. Methods Eng.* **2010**, *82*, 591–616. [CrossRef]
162. Yoon, G.H.; Jensen, J.S.; Sigmund, O. Topology optimization of acoustic–structure interaction problems using a mixed finite element formulation. *Int. J. Numer. Methods Eng.* **2007**, *70*, 1049–1075. [CrossRef]
163. Kreissl, S.; Pingen, G.; Evgrafov, A.; Maute, K. Topology optimization of flexible micro-fluidic devices. *Struct. Multidiscip. Optim.* **2010**, *42*, 495–516. [CrossRef]
164. Yoon, G.H. Topological layout design of electro-fluid-thermal-compliant actuator. *Comput. Methods Appl. Mech. Eng.* **2012**, *209–212*, 28–44. [CrossRef]
165. Yoon, G.H. Stress-based topology optimization method for steady-state fluid–structure interaction problems. *Comput. Methods Appl. Mech. Eng.* **2014**, *278*, 499–523. [CrossRef]
166. Yoon, G.H. Compliant topology optimization for planar passive flap micro valve. *J. Nanosci. Nanotechnol.* **2014**, *14*, 7585–7591. [CrossRef]
167. Jenkins, N.; Maute, K. An immersed boundary approach for shape and topology optimization of stationary fluid–structure interaction problems. *Struct. Multidiscip. Optim.* **2016**, *54*, 1191–1208. [CrossRef]

168. Munk, D.J.; Kipouros, T.; Vio, G.A.; Steven, G.P.; Parks, G.T. Topology optimisation of micro fluidic mixers considering fluid–structure interactions with a coupled Lattice Boltzmann algorithm. *J. Comput. Phys.* **2017**, *349*, 11–32. [CrossRef]
169. Picelli, R.; Vicente, W.M.; Pavanello, R. Evolutionary topology optimization for structural compliance minimization considering design-dependent FSI loads. *Finite Elem. Anal. Des.* **2017**, *135*, 44–55. [CrossRef]
170. Yoon, G.H. Brittle and ductile failure constraints of stress-based topology optimization method for fluid–structure interactions. *Comput. Math. Appl.* **2017**, *74*, 398–419. [CrossRef]
171. Lundgaard, C.; Alexandersen, J.; Zhou, M.; Andreasen, C.S.; Sigmund, O. Revisiting density-based topology optimization for fluid–structure interaction problems. *Struct. Multidiscip. Optim.* **2018**, *58*, 969–995. [CrossRef]
172. Munk, D.J.; Kipouros, T.; Vio, G.A.; Parks, G.T.; Steven, G.P. On the effect of fluid–structure interactions and choice of algorithm in multi-physics topology optimisation. *Finite Elem. Anal. Des.* **2018**, *145*, 32–54. [CrossRef]
173. Munk, D.J.; Kipouros, T.; Vio, G.A. Multi-physics bi-directional evolutionary topology optimization on GPU-architecture. *Eng. Comput.* **2018**, *35*, 1059–1079. [CrossRef]
174. Feppon, F.; Allaire, G.; Bordeu, F.; Cortial, J.; Dapogny, C. Shape optimization of a coupled thermal fluid–structure problem in a level set mesh evolution framework. *SeMA J.* **2019**, *76*, 413–458. [CrossRef]
175. Guest, J.K.; Prévost, J.H. Design of maximum permeability material structures. *Comput. Methods Appl. Mech. Eng.* **2007**, *196*, 1006–1017. [CrossRef]
176. Guest, J.K.; Prévost, J.H. Optimizing multifunctional materials: Design of microstructures for maximized stiffness and fluid permeability. *Int. J. Solids Struct.* **2006**, *43*, 7028–7047. [CrossRef]
177. Hollister, S.J.; Lin, C.Y. Computational design of tissue engineering scaffolds. *Comput. Methods Appl. Mech. Eng.* **2007**, *196*, 2991–2998. [CrossRef]
178. Xu, S.; Cheng, G. Optimum material design of minimum structural compliance under seepage constraint. *Struct. Multidiscip. Optim.* **2010**, *41*, 575–587. [CrossRef]
179. Andreasen, C.S.; Sigmund, O. Saturated poroelastic actuators generated by topology optimization. *Struct. Multidiscip. Optim.* **2010**, *43*, 693–706. [CrossRef]
180. Chen, Y.; Zhou, S.; Li, Q. Microstructure design of biodegradable scaffold and its effect on tissue regeneration. *Biomaterials* **2011**, *32*, 5003–5014. [CrossRef]
181. Chen, Y.; Schellekens, M.; Zhou, S.; Cadman, J.; Li, W.; Appleyard, R.; Li, Q. Design Optimization of Scaffold Microstructures Using Wall Shear Stress Criterion Towards Regulated Flow-Induced Erosion. *J. Biomech. Eng.* **2011**, *133*, 081008. [CrossRef]
182. Gonçalves Coelho, P.; Rui Fernandes, P.; Carriço Rodrigues, H. Multiscale modeling of bone tissue with surface and permeability control. *J. Biomech.* **2011**, *44*, 321–329. [CrossRef]
183. Challis, V.J.; Guest, J.K.; Grotowski, J.F.; Roberts, A.P. Computationally generated cross-property bounds for stiffness and fluid permeability using topology optimization. *Int. J. Solids Struct.* **2012**, *49*, 3397–3408. [CrossRef]
184. Andreasen, C.S.; Sigmund, O. Multiscale modeling and topology optimization of poroelastic actuators. *Smart Mater. Struct.* **2012**, *21*, 065005. [CrossRef]
185. Andreasen, C.S.; Sigmund, O. Topology optimization of fluid–structure-interaction problems in poroelasticity. *Comput. Methods Appl. Mech. Eng.* **2013**, *258*, 55–62. [CrossRef]
186. Youssef, K.; Mack, J.J.; Iruela-Arispe, M.L.; Bouchard, L.S. Macro-scale topology optimization for controlling internal shear stress in a porous scaffold bioreactor. *Biotechnol. Bioeng.* **2012**, *109*, 1844–1854. [CrossRef]
187. Ha, S.H.; Lee, H.Y.; Hemker, K.J.; Guest, J.K. Topology Optimization of Three-Dimensional Woven Materials Using a Ground Structure Design Variable Representation. *J. Mech. Des.* **2019**, *141*. [CrossRef]
188. Wein, F.; Chen, N.; Iqbal, N.; Stingl, M.; Avila, M. Topology optimization of unsaturated flows in multi-material porous media: Application to a simple diaper model. *Commun. Nonlinear Sci. Numer. Simul.* **2019**, *78*, 104871. [CrossRef]
189. Takezawa, A.; Zhang, X.; Tanaka, T.; Kitamura, M. Topology optimisation of a porous unit cell in a fluid flow considering Forchheimer drag. *Int. J. Comput. Fluid Dyn.* **2019**. [CrossRef]
190. Lurie, S.A.; Rabinskiy, L.N.; Solyaev, Y.O. Topology optimization of the wick geometry in a flat plate heat pipe. *Int. J. Heat Mass Transf.* **2019**, *128*, 239–247. [CrossRef]

191. Takezawa, A.; Zhang, X.; Kato, M.; Kitamura, M. Method to optimize an additively-manufactured functionally-graded lattice structure for effective liquid cooling. *Addit. Manuf.* **2019**, *28*, 285–298. [CrossRef]
192. Takezawa, A.; Zhang, X.; Kitamura, M. Optimization of an additively manufactured functionally graded lattice structure with liquid cooling considering structural performances. *Int. J. Heat Mass Transf.* **2019**, *143*, 118564. [CrossRef]
193. Bazilevs, Y.; Takizawa, K.; Tezduyar, T.E. New directions and challenging computations in fluid dynamics modeling with stabilized and multiscale methods. *Math. Models Methods Appl. Sci.* **2015**, *25*, 2217–2226. [CrossRef]
194. Bazilevs, Y.; Takizawa, K.; Tezduyar, T.E. Computational analysis methods for complex unsteady flow problems. *Math. Models Methods Appl. Sci.* **2019**, *29*, 825–838. [CrossRef]
195. Sigmund, O. On the usefulness of non-gradient approaches in topology optimization. *Struct. Multidiscip. Optim.* **2011**, *43*, 589–596. [CrossRef]
196. Alexandersen, J. Topology Optimisation for Coupled Convection Problems. Master's Thesis, Technical University of Denmark (DTU), Kongens Lyngby, Denmark, 2013.

© 2020 by the authors. Licensee MDPI, Basel, Switzerland. This article is an open access article distributed under the terms and conditions of the Creative Commons Attribution (CC BY) license (http://creativecommons.org/licenses/by/4.0/).

Article

Computational Optimization of Adaptive Hybrid Darrieus Turbine: Part 1

Palanisamy Mohan Kumar [1,*], Mohan Ram Surya [2], Krishnamoorthi Sivalingam [3,4], Teik-Cheng Lim [4], Seeram Ramakrishna [1] and He Wei [5]

1. Centre for Nanofibers and Nanotechnology, Department of Mechanical Engineering, National University of Singapore, Engineering Drive 3, Singapore 117587, Singapore; seeram@nus.edu.sg
2. Energy Research Institute, Nanyang Technological University, Innovation Centre, 71 Nanyang Drive, Singapore 638075, Singapore; mrsurya@ntu.edu.sg
3. NTUitive, Nanyang Technological University, Innovation Centre, 71 Nanyang Drive, Singapore 638075, Singapore; krishnamoorthi001@suss.edu.sg
4. School of Science and Technology, Singapore University of Social Sciences, 463 Clementi Rd, Singapore 59949, Singapore; tclim@suss.edu.sg
5. Singapore Institute of Manufacturing Technology, Surface Technology group, A*STAR, Fusionopolis way 2, Innovis, Singapore 138634, Singapore; hewei@simtech.a-star.edu.sg
* Correspondence: mohan@nus.edu.sg

Received: 17 April 2019; Accepted: 16 May 2019; Published: 17 May 2019

Abstract: Darrieus-type Vertical Axis Wind Turbines (VAWT) are promising for small scale decentralized power generation because of their unique advantages such as simple design, insensitive to wind direction, reliability, and ease of maintenance. Despite these positive aspects, poor self-starting capability and low efficiency in weak and unsteady winds deteriorate further development. Adaptive Hybrid Darrieus Turbine (AHDT) was proposed by the author in the past study as a potential solution to enhance low wind speed characteristics. The objective of the current research is to optimize the parameters of AHDT. AHDT integrates a dynamically varying Savonius rotor with a Darrieus rotor. A fully detailed 2D numerical study employing Reynold-Averaged Navier Stokes (RANS) is carried out to investigate the impact of the Darrieus rotor diameter (D_R) on the Savonius rotor (D_T) with regard to hybrid turbine performance. The power coefficient of the Darrieus rotor is evaluated when the Savonius rotor is in the closed condition (cylinder) of various diameters. The influence of Reynolds number (Re) on the torque coefficient is examined. Power loss of 58.3% and 25% is reported for D_R/D_T ratio of 1.5 and 2 respectively for AHDT with solidity 0.5 at 9 m/s. The flow interaction between the Savonius rotor in closed configuration reveals the formation of von Karman vortices that interact with Darrieus blades resulting in flow detachment. An optimum diametrical ratio (D_R/D_T) of 3 is found to yield the maximum power coefficient of the Darrieus rotor.

Keywords: wind turbine; Savonius; Darrieus; power coefficient; torque coefficient; wake

1. Introduction

The insatiable hunger for energy and the mass burning of fossil fuels has shifted the focus towards clean energy sources. Wind energy, being the key source among renewable energies, saw tremendous growth in the past decades with a total installed capacity of 591 GW as of 2018 [1]. The rapid development of wind turbine technology has sprouted interest in different wind turbines apart from Horizontal Axis Wind Turbine (HAWT). Advancements in energy storage technologies and decentralized power generation have renewed interest in VAWTs. They are especially suitable for small-scale power generation because of their niche advantages such as being insensitive to wind direction, ease of maintenance, reliability, aesthetics, and low noise. Among various types of

VAWT, Darrieus and Savonius types are popular as they are lucrative from an economic perspective. These turbines are especially suited for the urban environment and rooftop installations, as they are able to generate power in turbulent winds [2]. Savonius turbine operates by the difference in the aerodynamic drag between the concave and convex side of the buckets [3]. These turbines are able to start at low wind speed and generate substantial power. As per IEC 61400-2 [4], wind speed is considered to be low (Class IV) if the annual average wind speed at a specific location is below 7.5 m/s. Notable drawbacks include low efficiency and high wind loads demanding heavier support structures. Myriad studies have been carried out in the past, unfolding the flow pattern around the Savonius buckets and its blade wake interactions [5]. Numerous experimental studies established the critical parameters of the turbine such as the effect of end plates; optimum end plate diameters; the effect of aspect ratio; the influence of bucket spacing; bucket overlap; number of buckets; number of stages; interference of shat; and the effect of deflecting plate and helical and straight blades. Telescopic Savonius Turbine (TST) was proposed by Mohan [6] to avert the wind load at high winds. The numerical study concludes that the wind load can be reduced by up to 60% compared to the conventional Savonius turbine [7]. Darrieus turbine is known for its high efficiency among VAWTs as they operate by lift force, but they are notorious for their poor self-starting capability and low efficiency at weak wind flows. The helical turbine was proposed by Gorlov in 1995 [8]. Compared to the straight-bladed turbine, the helical-bladed turbine offers noteworthy advantages such as enhanced self-starting capability, low noise, increased blade life, low vibrations, and reduced peak stress in the blades [9]. The self-starting capability is improved by the reduction of Angle of Attack (AoA) and the ability of the blades to accelerate beyond the dead band [10].

It is also evident from the past studies that the stall angle is increased due to the boundary layer attachment by the introduction of spanwise flow by the helical blades. Tailored airfoils are proposed to delay stall [11]. Trapped vortex airfoil was conceived to delay the flow attachment by trapping a vortex bubble [12]. The numerical study reveals that the trapped bubble is unstable due to high AoA [13]. The J-profile blade is introduced to exploit the drag force and to reduce the blade manufacturing cost. Nested Darrieus rotor was attempted to increase the startup torque, but they induce vibrations on the downwind half due to turbulent wake downstream [14]. High solidity blades are able to increase the starting torque but the peak power coefficient is reduced significantly. Airfoils with a modified trailing edge were anticipated to provide an early start [15]. The cavity introduced on the trailing edge on the blade does improve the performance at low winds but suffers from reduced lift at high Tip Speed Ratio (TSR). Hybrid Savonius–Darrieus rotor demonstrates improved low wind behaviour but limits the operating TSR range to 1.2 on the combined machine [16]. The electrical startup was investigated to start the Darrieus rotor in low wind speed by accelerating beyond the dead band [17]. Innovative electronics convert the generator to a motor at low winds and operate as a conventional generator after reaching a critical rpm. Though the solution is attractive, the frequent starting due to intermittent wind consumes significant power leading to decreased annual energy yield. Blade pitching [18] was proposed to increase low wind performance as blades are constantly moved to optimum AoA. Despite improved overall performance, the blade pitching systems are not lucrative for a small wind turbine due to its complexity. Ducted Darrieus turbines have been experimented to accelerate the wind flow before it reaches the rotor [19]. The wind loads and cost of support structures hinders any further development on that front.

The Savonius turbines are known for their higher starting torque, while the Darrieus turbines are notorious for poor starting torque. Obviously, the curious question is how to combine both the rotors to supplement each other. At low wind speeds (<3 m/s), the combined machine demonstrates the characteristics of a Savonius rotor, while at high winds (>4 m/s), the combined machine will behave as a conventional Darrieus rotor. Two configurations are practically feasible to combine a Savonius and Darrieus rotor. The Savonius rotor can be placed either above or below the Darrieus rotor, or the Savonius rotor can be nested inside the Darrieus rotor. The placement of the Savonius rotor outside the Darrieus rotor tends to increase the length of the rotating shaft, which in turn will lead to vibrations.

Hence, the pragmatic method is to nest the Savonius rotor within the Darrieus rotor. Based on the above idea, a number of past studies [20] have been conducted both experimentally and computationally. The objective of all these studies is to optimize the Savonius rotor diameter for maximum low wind performance and to maintain the C_p of the Darrieus rotor at high winds by minimizing the effect of the Savonius rotor [21]. The results show that the starting torque of the combined machine is much higher than the Darrieus rotor, while the peak C_p is achieved at lower TSR. The maximum operating TSR for the combined machine is 1–1.2 [22]. It has to be noted that the maximum TSR for a Savonius rotor is 1, and any attempt to operate beyond TSR 1 will generate negative torque. The Darrieus rotor operates at TSR 3–6. This mismatch between the peak performance of two rotors is a crucial problem limiting the further development of these hybrid turbines [23]. All the concepts so far discussed are successful in improving the low wind behaviour to some extent, but severely degrade the performance at high winds. Innovative AHDT is proposed as an economically feasible solution with an early start and minimal impact on the Darrieus rotor by exploiting the merits of both the turbines.

2. State of art of Computational Simulations on VAWT

The renewed interest in the Darrieus turbines has resulted in intense research which has led to further understanding of the behaviors of the rotor. The dynamic stall has been studied in detail by Buchner by comparing the simulation and experiment. Simao [24] conducted a Particle Velocity Interferometry (PIV) visualization during the dynamic stall on the rotor. Flow curvature effects and angle of incidence was studied by Bianchini [25]. Blade wake interactions and the unsteady effects were investigated by Ragni [26]. High fidelity CFD simulations on the Darrieus rotor were performed by Balduzzi [27]. The effect of pitch angle on the performance of the Darrieus rotor and the associated aerodynamics were examined by Kalkman [28]. New airfoils have been proposed to lower the manufacturing cost such as NTU-20-V and can computationally be compared with NACA 0018 [29]. A similar study on the performance of the Darrieus turbine under different shaft diameters was carried out by Rezaeiha [30].

3. Overview of Adaptive Hybrid Darrieus Turbine

A hybrid Savonius–Darrieus turbine was proposed in the 1980s, as the two-bladed phi rotor suffers from poor self-starting capability. Another important challenge that hinders the development of the Darrieus rotor is the over-speeding of the turbine. Several FloWind turbines crashed to the ground due to inability of the turbines to limit their speed. The catastrophic failures of Flowind turbines (FloWind Corp., Tehachapi, CA, USA) has brought about deep concerns regarding the suitability of the Darrieus rotor for wind power generation. How to overcome poor self-starting capability and over-speeding are the two critical questions that have remained unanswered for these many years. A single solution to the above said questions conflicts each other, as enhancing the startup capability at a low wind speed will eventually lead to over-speeding at high winds. Though a Hybrid Savonius–Darrieus turbine is a potential solution to increase the starting torque, it drastically increases the wind loads and the over-speeding issues. The conventional hybrid turbine has a common shaft for Savonius and Darrieus rotors. The strategy is to employ the drag torque generated by a Savonius rotor to accelerate the Darrieus rotor. The optimum TSR for a two-bladed Darrieus rotor lies between 3 to 5 and the optimum TSR for Savonius rotor is 1 [31]. The Savonius rotor tends to generate resistive torque and in fact energy must be expended to rotate the Savonius rotor for a TSR above 1. The mismatch between the optimum TSR for the two rotors severely degrades the performance at higher TSR. Practically, a conventional hybrid Darrius-Savonius rotor will not accelerate beyond 1.5, resulting in an iota of improvement in annual energy output. Hence, a novel design has been put forward to minimize the influence of the Savonius rotor beyond a TSR of 1. The strategy is to transform the Savonius rotor into a shape that leaves minimum wake downstream without any resistive torque at higher TSR. A two bucket Savonius rotor can be transformed into a nominal cylinder if they are able to slide. The wake behind the downstream is axisymmetric with minimum width compared to other shapes. The wake width and the kinetic energy imbibed dictate the performance of the Darrieus rotor. A two-stage two bucket Savonius rotor offset at 90° can improve the directional starting. The three operating configurations are shown in Figure 1a–c. At low wind speed, the Darrieus rotor torque ($M_d = +ve$) and Savonius rotor torque ($M_s = +ve$) are in the same direction when the Savonius buckets a and b are arranged as shown in Figure 1a.

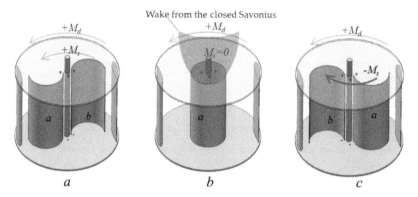

Figure 1. (a) Adaptive Hybrid Darrieus Turbine (AHDT) rotor in open configuration (b) Closed configuration (c) Braking configuration.

As the TSR reaches 1, the buckets slide towards the axis of rotation to form a cylinder without any torque generation ($M_s = 0$), as shown in Figure 1b. In the extreme wind conditions and above the rated rotor rpm, the Savonius buckets slides in the opposite direction, generating resistive torque ($M_s = -ve$), decelerating the rotor as shown in Figure 1c. The number of blades on the Darrieus rotor, orientation of the rotor to the oncoming wind, angular offset between the Savonius buckets and the Darrieus rotor, and ratio between the diameters of Darrieus rotor to the Savonius rotor are the crucial design parameters that determine the starting characteristics. The angular offset between Darrieus blades and the Savonius buckets will have minimal impact on the performance, as two stages are arranged offset at 90°. Hence, from the structural perspective, the Savonius buckets can slide on the Darrieus blade, connecting struts and eliminating the requirement for additional structures. Thus, the AHDT has the capability to start the turbine at low wind speed, let it operate with minimal effect on the Darrieus rotor, and decelerate the rotor when it rotates beyond the rated rpm. The construction and mechanical arrangement are less complex making this concept commercially implementable. A double Multiple StreamTube model [32] employed for the performance assessment of the conventional Darrieus rotor is not applicable to predict the performance of AHDT, though the model was modified to include the Savonius rotor [33]. Hence, computational analysis will be a suitable process to initiate the optimization with the initial promising experimental results [34].

4. Computational Domain and Meshing

The guidelines for the computational simulations for this work are derived from a previous study [35]. It is vital to consider a larger domain size that represents the turbine operating in field conditions and to avoid the blockage. The computational domain is rectangular and of the size 20 D × 35 D, where D corresponds to the maximum rotor diameter (0.4 m in current study), and the rotor is placed at 10 D from the inlet boundary condition. In addition, the minimum domain size in all directions should be at least 20 times the rotor radius. The exit boundary condition is placed at a distance of 15 D downstream with respect to the rotor in order to allow a complete development of the wake structure. The inlet boundary has been set as a velocity inlet, while the outlet is set to a pressure outlet. The wind speed at the inlet varies from 4 m/s to 9 m/s. The pressure outlet has been assigned to an atmospheric pressure value. For the two side boundaries, symmetric boundary conditions have been used. To maintain the flow continuity in the field, the circular edge in the domain is set as an interface. In order to maintain the mesh linkage between the domain and the turbine, the circumferential edge of the turbine is also set as an interface. The rotating zone revolves at the same angular speed as the rotor, while the domain region is stationary. Since the flow is over a rotating solid, the sliding mesh technique has been used. Figure 2a shows the main geometric features and the CFD boundary conditions of the computational domain. A 2D unstructured mesh (triangular type) has been generated in the domain as well as in the rotating zone using ANSYS meshing tool as shown in Figure 2b,c. Much attention has been paid for the near wall treatment of the blades by generating cells of smaller sizes with an edge size setting of 0.1 mm. From the blade walls, the mesh cells grow evenly from smaller size topology to higher ones as seen in the Figure 3, thus adopting the appropriate cell size prevailing in the rotating zone. In the vicinity of the blade, structured mesh has been generated with y+ < 1 to capture the flow near the blades accurately.

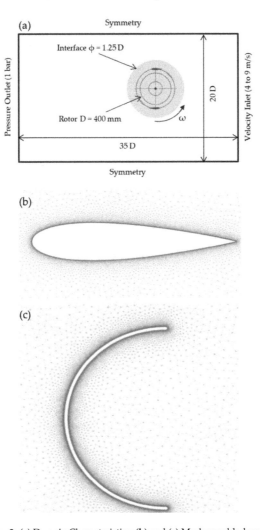

Figure 2. (a) Domain Characteristics; (b) and (c) Mesh near blade walls.

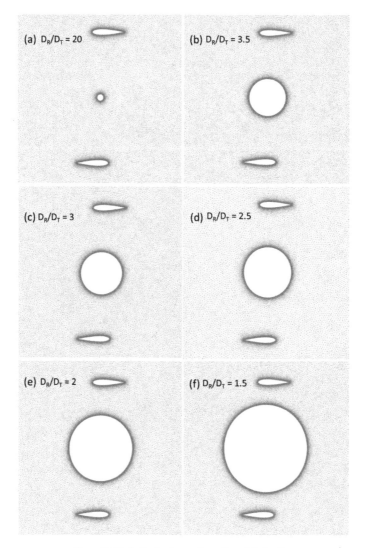

Figure 3. Mesh display for various diametrical ratios.

5. Convergence Analysis

5.1. Grid Convergence Analysis

The computational results are highly dependent on the density of the grid. Mesh-independent results are imperative for accurate prediction and to compare with experimental outcomes. Coarse, medium and fine grids are generated with the cell count of 0.3, 0.5 and 0.8 million respectively. Though the conventional method is to systematically double the number of cells following established methods such as grid systematic refinement, General Richardson Extrapolation (GRE) or Grid Convergence Index (GCI), the current study relies on prior similar studies of cell count to start with in order to reduce the computational time. The number of nodes on the airfoil surface is 1000 for the coarse grid, whereas for the fine grid it is improved to 4000 nodes to capture the complex flow, flow detachment and dynamic stall. The torque coefficient (Ct) is the primary parameter through which the power coefficient is obtained. It is defined as the ratio of generated aerodynamic torque to the available torque in the wind. Ct is computed by applying the RANS and the SST turbulence model for the selected grid sizes. Results depict that the difference in Ct value between 0.5 and 0.8 million cells is less than 1%. The same procedure is repeated for the two-bladed and three-bladed Darrieus rotors to confirm that the chosen mesh density is sufficient for a solidity of $\sigma = 0.5$ and $\sigma = 0.75$. Hence, the optimum cell count is concluded as 0.5 million, which will be followed for the rest of study. The Ct value against different cell counts is shown in Figure 4a.

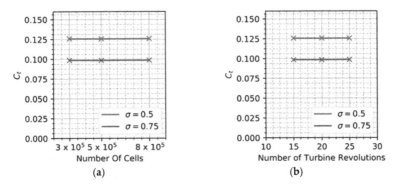

Figure 4. (a) Grid independence study on different solidity. (b) Time step on different solidity.

5.2. Revolution Convergence Analysis

As the turbine starts to rotate, the flow over the turbine blades is highly unstable. After a few revolutions, the flow is stabilized at which the parameter of interest is measured. Finding the minimum number of revolutions for the given domain is imperative to keep the computational resources and time low. The number of revolutions at which the flow stabilizes is dictated by domain size, number of cells and boundary conditions. For the current test case, the Ct value is monitored over 30 revolutions. The turbine is allowed to rotate at a different TSR from 0.5 to 4, with the turbulent intensity and turbulent length scale fixed. Segregated pressure-based solver, known as a SIMPLE (Semi Implicit Method for Pressure Linked Equations) algorithm, is employed to execute the computation. The computation is initiated with a first order upwind scheme for spatial and temporal discretization followed by a second-order upwind scheme for accurate predictions. The time step size is set to 5×10^{-4} s. Residual convergence criteria for each physical step is set to 10^{-5}. The Ct is measured for each timestep and averaged over one revolution. The obtained Ct value is compared with the previous revolution to compute the difference and corresponding number of revolutions chosen if the difference is less than 0.02. The current run stabilizes after 20 revolutions and hence for further test cases, all the parameters of investigation are measured at 20 revolutions. The Ct value against the number of revolutions is shown in Figure 4b.

6. Mathematical Methodology

6.1. Reynolds Averaged Navier–Stokes Model

The Reynold-Averaged Navier–Stokes (RANS) method is the fastest CFD approach widely employed to solve the flow problems in wind turbines. The RANS models apply the eddy-viscosity model or Reynolds stress model to compute this Reynolds stress variable. Two equation models such as k-epsilon (k-ε), k-omega (k-ω) and k-ω SST are most widely used for wind turbines, where k is the turbulence kinetic energy per unit mass, epsilon is the specific dissipation, and ω is the specific dissipation rate. As mentioned before, the Navier–Stokes equations are very complex to solve directly due to their non-linear nature. This is due to the presence of convective acceleration terms. In order to approximate these equations, several numerical methods are available.

$$\rho \bar{u}_i \frac{\partial \bar{u}_i}{\partial x_j} = \rho \bar{f}_i + \frac{\partial}{\partial x_i}\left[-\bar{p}\delta_{ij} + \mu\left(\frac{\partial \bar{u}_i}{\partial x_j} + \frac{\partial \bar{u}_j}{\partial x_i}\right) - \rho \overline{u'_i u'_j}\right] \tag{1}$$

6.2. Turbulence Model

The k-ω SST model is a simpler way of representing the k-ω shear stress transport model which is the blend of k-ω and k-ε models. The k-ω SST model is the upgraded form of the baseline (BSL) model which varies linearly between k-ω and k-ε models. This model makes use of the k-ω definition in regions where the boundary layer predominates and the k-ε definition in regions outside the boundary layer. The transformed k-ω formulation from the k-ε model is given by

$$\rho \frac{\partial k}{\partial t} + \rho u_j \frac{\partial k}{\partial x_j} = \tau_{ij}\frac{\partial u_j}{\partial x_j} - \beta_1^{sst}\rho k\omega + \frac{\partial}{\partial x_j}\left[(\mu + \sigma_1^{sst}\mu_t)\frac{\partial k}{\partial x_j}\right] \tag{2}$$

$$\rho \frac{\partial \omega}{\partial t} + \rho u_j \frac{\partial \omega}{\partial x_j} = a^{sst}\frac{\omega}{k}\tau_{ij}\frac{\partial u_j}{\partial x_j} - \beta_2^{sst}\rho\omega^2 + \frac{\partial}{\partial x_j}\left[(\mu + \sigma_2^{sst}\mu_t)\frac{\partial \omega}{\partial x_j}\right] + 2\rho(1-F_1)\sigma_2^{sst}\frac{1}{\omega}\frac{\partial k}{\partial x_j}\frac{\partial \omega}{\partial x_j} \tag{3}$$

The newly transformed equations due to the combination of two models are represented by

$$\phi_{sst} = F_1\phi_\omega + (1-F_1)\phi_\epsilon \tag{4}$$

The blending function F1 switches between the previously defined two methods in the desired regions. It is formulated in such a way that the near wake region marked the use of the k-ω model and the free shear layer makes use of the k-ε model. Equations (5) and (6) mention that the F1 term represents the blending function which makes use of both fluid and special terms,

$$F_1 = \tanh\left(\left(\min\left(\max\left(\frac{\sqrt{k}}{0.09\omega y_s}, \frac{500v}{y_s^2\omega}\right), \frac{4\rho\sigma_2^{\omega\epsilon}}{CD_{k\epsilon}y_s^2}\right)\right)^4\right) \tag{5}$$

$$CD_{k\omega} = \max\left(2\rho\sigma_2^{\omega\epsilon}\frac{1}{x}\frac{\partial k}{\partial x_j}\frac{\partial \omega}{\partial x_j}, 10^{-10}\right) \tag{6}$$

The three factors mentioned in the above Equations (5)–(7) represent the viscous sub-layer, turbulent length scale and small free stream values. One of the major advantages of the k-ω SST model in comparison with the BSL model is that the turbulent shear stress is accounted for by limiting eddy viscosity. The turbulent shear stress is assumed to be proportional to specific turbulent kinetic energy in logarithmic and wake regions of the turbulent boundary layer. The major drawback of eddy viscosity and the Boussinesq hypothesis is that these models assume isotropic turbulence which would end up generating unrealistic results. k-ω SST is suitable for predicting the flow in the viscous sub-layer and in the regions away from the wall (wake region). In addition, the SST model is less sensitive to free stream conditions (flow outside the boundary layer) than many other turbulence models.

$$\mu_T = \frac{\rho a_1 k}{\max(a_1\omega, SF_2)} \tag{7}$$

$$F_2 = \tanh\left(\left(\max\left(\frac{\sqrt{k}}{0.09\omega y_s}, \frac{500v}{y_s^2\omega}\right)\right)^2\right) \tag{8}$$

7. Results and Discussion

7.1. Torque Coefficient Comparison for Different (D_R/D_T)

CFD simulations for different Darrieus to closed Savonius rotor diameter ratios (D_R/D_T) are performed in search of the optimum rotor diameters. The AHDT in closed configuration is simulated for five complete revolutions instantaneously recording the torque coefficient of blades for every degree of rotation. However, the solution is converged after three revolutions of the turbine. The coefficient of power (C_p) is calculated from (C_t) as shown in Equation (9).

$$C_p = C_t \times TSR \tag{9}$$

The operating TSR range (λ) considered in the present work is 1 to 4 for the two-bladed rotor at an increment of 0.2. Computing at a TSR, below the mentioned range, is of no significance for this optimization study, and above the indicated range means operating beyond stall conditions. For high solidities, power and efficiency rapidly decrease at stall conditions. The free stream wind speed for which the AHDT configurations are simulated is between 4 m/s and 9 m/s which is converted to Re for further descriptions in this study. The instantaneous torque coefficient (C_t) versus azimuth angle (θ) is computed for a range of Re = 1.2×10^5 to Re = 2.7×10^5. The results for the single blade are plotted in Figure 5a. For each Re, the various D_R/D_T ratios are evaluated for one complete revolution (0° to 360°). It shows that as D_R/D_T ratio increases, the peak torque coefficient in the C_t cycle increases. The torque coefficient for the rotor at different Re and different D_R/D_T is displayed in Figure 6. It can be seen that the two peak torque coefficients constantly decrease as the diameter of the cylinder increases. Also, it is evident that as the diameter of the cylinder increases, the azimuthal angle at peak C_t shifts. This can be attributed to the change in the relative wind velocity due to vortices from the cylinder.

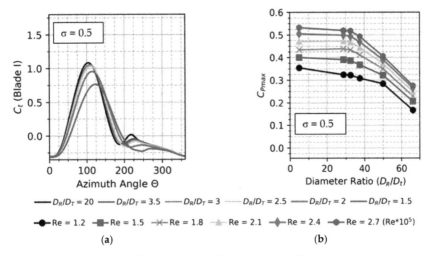

Figure 5. (a) Torque coefficient of single blade; (b) C_{pmax} for different diametrical ratio.

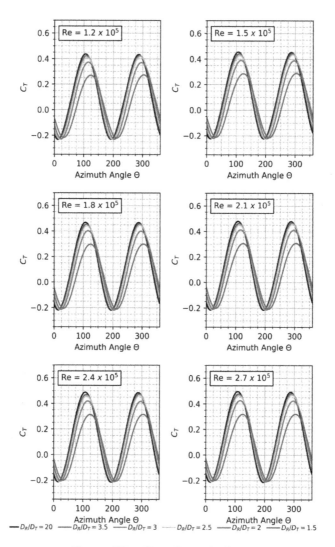

Figure 6. Effect of Re on Torque Coefficient.

7.2. Power Coefficient Comparison for Different (D_R/D_T)

Figure 7 shows the power curves of the AHDT for the two-bladed rotors corresponding to the solidity of $\sigma = 0.5$. Each subplot in Figure 7 shows the power curve (C_p vs. λ) for their respective diameter ratio (D_R/D_T). The power curves are calculated for a range of Re = 1.2×10^5 to Re = 2.7×10^5. Figure 5b shows the comparison of the maximum power coefficient (C_{pmax}) with the investigated Re and D_R/D_T ratio. As mentioned before, the D_R/D_T = 20 corresponds to the conventional Darrieus rotor. Figure 8 shows the C_p curve at different Re ranging from Re = 1.2×10^5 to Re = 2.7×10^5. It can be seen from Figure 8 that for the low Re of 1.2, 1.5 and 1.8×10^5, the optimal TSR is 3.2.

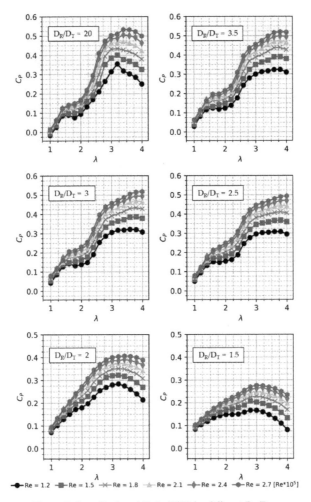

Figure 7. C_p vs Tip Speed Ratio (TSR) for different D_R/D_T.

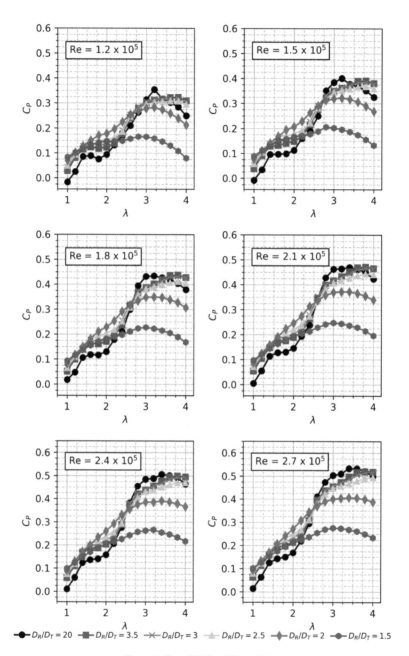

Figure 8. C_p vs TSR for different Re.

However, for a higher Re of 2.1, 2.4 and 2.7 × 10^5, the optimal TSR shifts to 3.4. The maximum power coefficient increases by 11.7% at Re = 1.5 × 10^5 when compared to Re = 1.2 × 10^5. At Re = 1.8, 2.1 and 2.4 × 10^5, the percentage increase in C_{pmax} is 18.4%, 24.6% and 30% higher than the C_{pmax} at Re = 1.2 × 10^5. At Re = 2.7 × 10^5, the C_{pmax} value is 0.533 which is 33.6% higher than the C_{pmax} obtained at the low Re of 1.2 × 10^5. For the diameter ratio up to D_R/D_T = 2.5, the C_p curves show a dip before reaching the maximum C_p, except for the $D_R/D_T = 20$. By poring over the data, it is evident that the power loss occurs at TSR ~2. The loss in power reflecting as the dip is attributed to the vortices from the cylinder interacting with the Darrieus blades. The frequency of the vortices can be deduced from Strouhal number (S_t) as given by Equation (10),

$$S_t = f_s D/V \qquad (10)$$

where f_s is the vortex-shedding frequency, D is the across-wind dimension of the body, and V is the mean velocity of the uniform flow. The frequency of the vortices and the blade passing frequency are correlated at TSR 2, resulting in power loss. The C_p curve of the larger diameter cylinder does not exhibit this kind of power loss, as the cylinder acts as a bluff body generating a large wake, where the wake width occupies most of the downstream path. For $D_R/D_T = 1.5$, the maximum C_p achieved for the Re = 2.7×10^5 is 0.27, whereas for the same Re, $D_R/D_T = 20$, the peak C_p is 0.53 (Table 1). It is evident that the diametrical ratio of 1.5 and 2 significantly reduce the power coefficient by more than half. Hence, these diameters are not suitable for AHDT. Another interesting finding is that the maximum C_p is a sharp curve for the lower diameter ratio, and as the cylinder diameter increases, the maximum C_p although it is low, is maintained over a wider TSR, making the curve flat rather than sharp. The advantage is that a higher power can be extracted from the turbine for a wide wind speed range.

Table 1. Difference in C_p at various Re.

Re × 10^5	D_R/D_T										
	20		3.5		3		2.5		2		1.5	
	C_p	Δ %	C_p	Δ %	C_p	Δ %	C_p	Δ %	C_p	Δ%	C_p	Δ%
1.2	0.35	-	0.32	-	0.32	-	0.30	-	0.24	-	0.10	-
1.5	0.40	11.7	0.39	17.1	0.38	17.3	0.36	16.3	0.29	17.2	0.15	30.6
1.8	0.43	18.4	0.43	26.0	0.43	26.1	0.41	25.1	0.32	26.8	0.19	42.9
2.1	0.47	24.6	0.47	31.3	0.46	36.6	0.44	30.6	0.35	32.6	0.21	49.5
2.4	0.50	30.0	0.49	35.1	0.49	35.2	0.47	34.5	0.36	34.6	0.23	54.2

The difference in the C_p values for a low Re of 1.2×10^5 is comparatively lower than the difference in C_p for the higher Re of 2.7×10^5 for different diameter ratios. At lower Re, the Darrieus rotor by itself has a lower C_p value and the impact of turbulent flow from the cylinder is comparatively low. The peak C_p value for Re 1.2×10^5 stays at 0.3 for all the investigated diameters. As the diametrical ratio increases to 1.5, the C_p curve stays at 0.1 for all the TSR. It can be concluded from the C_p curves comparison for various D_R/D_T that the $D_R/D_T = 2.5$, 2 and 1.5 reduces the C_p by more than half, hence these diameters are not suitable for AHDT further optimization.

7.3. Effect of Re on C_p for Different D_R/D_T

Re is a critical factor in determining the power production capability and starting behaviour of a Darrieus rotor. For a conventional Darrieus rotor, the starting torque entirely rests on the lift generated at low Re, but for AHDT, the starting torque is generated by drag rather than lift. Hence, the study on the influence of Re aims to shed light on the power performance of AHDT and flow over cylinder rather than the starting behaviour of AHDT at low Re. By comparing different diametrical ratios for the given Re, i.e. wind speed, a suitable cylinder diameter or in other words, the diameter of the Savonius buckets, can be determined.

It is not always pertinent to choose a smaller diameter cylinder to maximize the C_p of Darrieus rotor, as the vortices of the cylinder at a given Re can deteriorate the torque of Darrieus blades more than larger diameter cylinder. Hence, it is imperative to perform the simulation iteratively for $D_R/D_T = 20$ to $D_R/D_T = 1.5$. The results are critical from the structural perspective, as a loss in C_p will indicate a vortex shedding and further instigate to optimize the aspect ratio of the cylinder. Based on the Re regime, the flow over the cylinder may differ at large. The pressure gradient on the cylinder and the boundary layer may give rise to vortices of different diameters. The boundary layer upstream of the cylinder must overcome a strong pressure gradient set up by the cylinder. This leads to separation of the flow, and in the separated region, a vortex system is developed which is stretched around the cylinder like a horseshoe. Hence, AHDT behaviour at higher Re = 2.1×10^5 is of particular interest. The C_p curves for the investigated Re are shown in Figure 8. For all the diametrical ratios, except $D_R/D_T = 1.5$, the C_p curves are almost similar. The C_p values increase steadily as the Re increases. For 2.4×10^5, the peak C_p achieved is 0.3 for all the diametrical ratios except 1.5, for which the C_p is 0.2, due to wake expansion as seen from Figure 9. Vorticity contours corresponding to the $D_R/D_T = 1.5$ reveal that the Darrieus blades are operating in the large wake generated by the cylinder as shown in Figure 10.

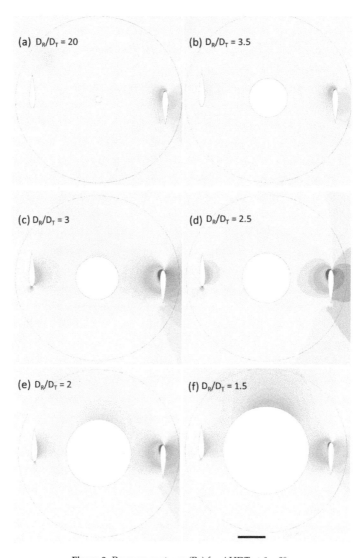

Figure 9. Pressure contours (Pa) for AHDT at $\beta = 0°$.

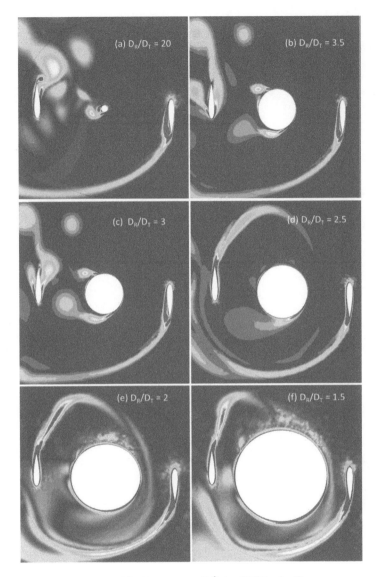

Figure 10. Vorticity contours (S^{-1}) for AHDT at $\beta = 0°$.

7.4. Discussion of Pressure and Vorticity Contours

The pressure contours for different diametrical ratios are compared in Figures 11 and 12 for azimuthal position 0° and 30° respectively. The azimuthal angle is of particular interest to analyze in detail, as the wake from the Darrieus blades and the Savonius buckets in closed conditions (cylinder) will be maximum, resulting in a higher power loss. For the rest of the azimuthal angles, the wake from the Darrieus blades are dispersed before it reaches the Savonius buckets. For the ratio $D_R/D_T = 20$, the flow pattern in the rotor is similar to a conventional Darrieus rotor, as the cylinder resembles the center shaft of the conventional turbine. For some of the cantilevered tower designs, the centre shaft diameters will be even higher than the diametrical ratio of 20. The wake pattern observed is similar to the Darrieus rotor, with low pressure behind the Darrieus blades with a peak power coefficient of 0.35. For $D_R/D_T = 3.5$, the cylinder wake starts to appear with a low-pressure zone downstream. The corresponding vorticity for the azimuthal angle $\beta = 0°$ indicates that the flow on the cylinder starts to separate with alternate vortices on both sides. The vortices are small and smoothly travel downstream, eventually regaining the freestream velocity. At $\beta = 30°$, a similar pressure drop occurs without any noticeable difference between two angles. For the ratio $D_R/D_T = 3$, the cylinder diameter is comparatively larger, and the pressure drops around the cylinder is noticeable.

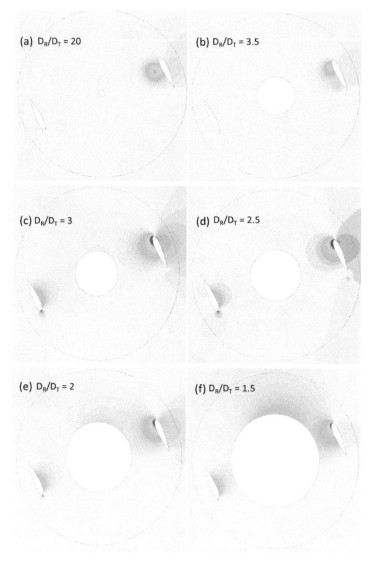

Figure 11. Pressure contours (Pa) for AHDT at $\beta = 30°$.

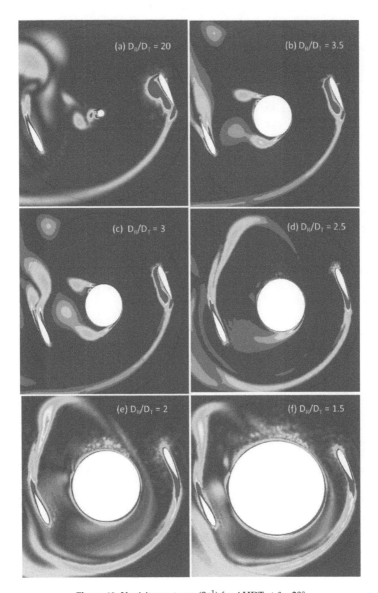

Figure 12. Vorticity contours (S^{-1}) for AHDT at $\beta = 30°$.

The wake arising due to the pressure difference from the Darrieus blades progressively extends to the center cylinder, forming a large low-pressure zone. Though the intensity is not high, the low-pressure zone influences the power loss, decreasing the peak power coefficient as evident from Figure 7. For the ratio D_R/D_T = 2.5, the cylinder influence on the power coefficient of the Darrieus rotor is well pronounced. From the pressure contours on Figure 11d, the low-pressure zone extends from the upstream Darrieus blades through the cylinder to the downstream blades. The significant loss in power can be attributed to the size of the vortices compared to the blade chord and the downstream path where it encounters the blades. TSR plays a crucial role in generating asymmetric alternating wake. Von Karman vortices are formed due to flow separation from the cylinder. Up to this diametrical ratio, the wake can be clearly distinguished between the Darrieus blades and the cylinder wake. As the cylinder diameter increases further to the ratio of $D_R/D_T = 2$, the whole rotor is in the wake, which is evident from the pressure contour as shown in Figure 11e,f. An interesting finding is that the flow starts to deflect before it reaches the cylinder, disturbing the flow upstream on the Darrieus rotor. Due to the low speed and high energy turbulent flow, the Darrieus blades also leave a large wake behind. As evident from the corresponding C_p graph, the power loss is significantly low as power generation happens only at limited azimuthal angles and the energy has to be expended for the majority of the downstream travel path. The same flow pattern repeats for other azimuthal positions. The von Karman vortices are unstable and break down due to Strouhal instability.

8. Conclusions

2D steady-state simulations are performed on the proposed AHDT in an effort to determine the optimum Savonius diameter that can be integrated with the Darrieus rotor to maximize its performance for the complete operating range. Turbulence is modelled with the k-ω SST equation. As the flow pattern over the cylinder will be sub critical or critical based on the Re, the study systematically investigates their performance on Darrieus rotor with a fixed diameter. The power loss of the Darrieus rotor is more than half for the cylinder diameter ratio (D_R/D_T) of 2 and 1.5, as power has to be expended for the blades when it passes through downstream. A smaller (D_R/D_T) leads to a smaller Savonius bucket diameter which reduces the turbine performance in the low wind speeds. Hence, it can be concluded that D_R/D_T should lie around 3 to maximize the Darrieus turbine performance for the whole of the wind speed spectrum. The optimum diameter can be concluded after evaluating the performance of the high solidity three-bladed turbine and the starting capability when the Savonius rotor is in open condition. The starting performance of AHDT with open Savonius will be investigated in part 2, extending the current study. The investigation of the performance of AHDT with various airfoil profiles, solidity, and aspect ratios can be intriguing for future research.

Author Contributions: Funding acquisition, S.R. and H.W.; Project administration, H.W.; Supervision, T.-C.L.; Writing—original draft, P.M.K. and M.R.S.; Writing—review & editing, K.S.

Funding: This research was funded by 2016 ASTAR AME IRG Grant.

Conflicts of Interest: The authors declare no conflict of interest.

References

1. International Energy Agency. Available online: https://www.iea.org/ (accessed on 4 April 2019).
2. Karthikeya, B.; Negi, P.S.; Srikanth, N. Wind resource assessment for urban renewable energy application in Singapore. *Renew. Energy* **2016**, *87*, 403–414. [CrossRef]
3. Shikha; Bhatti, T.S.; Kothari, D.P. Early Development of Modern Vertical and Horizontal Axis Wind Turbines: A Review. *Wind Eng.* **2005**, *29*, 287–299. [CrossRef]
4. IEC 61400-2:2013 Wind Turbines Part 2: Small Wind Turbines. Available online: https://webstore.iec.ch/publication/5433 (accessed on 17 May 2019).
5. Golecha, K.; Kamoji, M.A.; Kedare, S.B.; Prabhu, S.V. Review on Savonius Rotor for Harnessing Wind Energy. *Wind Eng.* **2012**, *36*, 605–645. [CrossRef]
6. Mohan Kumar, P.; Surya, M.M.R.; Narasimalu, S.; Lim, T.-C. Experimental and numerical investigation of novel Savonius wind turbine. *Wind Eng.* **2019**, *43*, 247–262. [CrossRef]
7. Kumar, P.M.; Purimitla, S.R.; Shubhra, S.; Srikanth, N. Numerical and analytical study on telescopic savonius turbine blade. In Proceedings of the 2017 3rd International Conference on Power Generation Systems and Renewable Energy Technologies (PGSRET), Johor Bahru, Malaysia, 4–6 April 2017; IEEE: Piscataway, NJ, USA, 2017; pp. 107–112.
8. Gorlov, A. *Development of the Helical Reaction Hydraulic Turbine*; Final Technical Report; Northeastern Univ.: Boston, MA, USA, 1998.
9. Baker, J. Features to aid or enable self starting of fixed pitch low solidity vertical axis wind turbines. *J. Wind Eng. Ind. Aerodyn.* **1983**, *15*, 369–380. [CrossRef]

10. Shiono, M.; Suzuki, K.; Kiho, S. Output characteristics of Darrieus water turbine with helical blades for tidal current generations. In Proceedings of the Twelfth International Offshore and Polar Engineering Conference, Kitakyushu, Japan, 26–31 May 2002; ISOPE: Cupertino, CA, USA, 2002.
11. Kumar, M.; Surya, M.M.R.; Sin, N.P.; Srikanth, N. Design and experimental investigation of airfoil for extruded blades. *Int. J. Adv. Agric. Environ. Eng.* **2017**, *3*, 359–400.
12. Kumar, P.M.; Surya, M.M.R.; Kethala, R.; Srikanth, N. Experimental investigation of the performance of darrieus wind turbine with trapped vortex airfoil. In Proceedings of the 2017 3rd International Conference on Power Generation Systems and Renewable Energy Technologies (PGSRET), Johor Bahru, Malaysia, 4–6 April 2017; IEEE: Piscataway, NJ, USA, 2017; pp. 130–135.
13. Kumar, P.M.; Surya, M.M.R.; Srikanth, N. On the improvement of starting torque of darrieus wind turbine with trapped vortex airfoil. In Proceedings of the 2017 IEEE International Conference on Smart Grid and Smart Cities (ICSGSC), Singapore, 23–26 July 2017; IEEE: Piscataway, NJ, USA, 2017; pp. 120–125.
14. Kumar, P.M.; Ajit, K.R.; Surya, M.R.; Srikanth, N.; Lim, T.-C. On the self starting of darrieus turbine: An experimental investigation with secondary rotor. In Proceedings of the 2017 Asian Conference on Energy, Power and Transportation Electrification (ACEPT), Singapore, 24–26 October 2017; IEEE: Piscataway, NJ, USA, 2017; pp. 1–7.
15. Kumar, P.M.; Kulkarni, R.; Srikanth, N.; Lim, T.-C. Performance Assessment of Darrieus Turbine with Modified Trailing Edge Airfoil for Low Wind Speeds. *Smart Grid Renew. Energy* **2017**, *8*, 425–439. [CrossRef]
16. Wakui, T.; Tanzawa, Y.; Hashizume, T.; Nagao, T. Hybrid configuration of Darrieus and Savonius rotors for stand-alone wind turbine-generator systems. *Electr. Eng. Jpn.* **2005**, *150*, 13–22. [CrossRef]
17. Kjellin, J.; Bernhoff, H. Electrical starter system for an H-rotor type VAWT with PM-generator and auxiliary winding. *Wind Eng.* **2011**, *35*, 85–92. [CrossRef]
18. Chougule, P.D.; Nielsen, S.R.K.; Basu, B. Active Blade Pitch Control for Straight Bladed Darrieus Vertical Axis Wind Turbine of New Design. *Key Eng. Mater.* **2013**, *569–570*, 668–675. [CrossRef]
19. Hwang, I.S.; Min, S.Y.; Jeong, I.O.; Lee, Y.H.; Kim, S.J. Efficiency improvement of a new vertical axis wind turbine by individual active control of blade motion. In Proceedings of the SPIE Smart Structures and Materials + Nondestructive Evaluation and Health Monitoring, San Diego, CA, USA, 5 April 2006; SPIE: Bellingham, WA, USA; p. 617311.
20. Rassoulinejad-Mousavi, S.; Jamil, M.; Layeghi, M. Experimental study of a combined three bucket H-rotor with savonius wind turbine. *World Appl. Sci. J.* **2013**, *28*, 205–211.
21. Gupta, R.; Das, R.; Sharma, K. Experimental study of a Savonius-Darrieus wind machine. In Proceedings of the International Conference on Renewable Energy for Developing Countries, Washington, DC, USA, 6 April 2006.
22. Gupta, R.; Sharma, K. Flow physics of a combined Darrieus-Savonius rotor using computational fluid dynamics (CFD). *Int. Res. J. Eng. Sci. Technol. Innov.* **2012**, *1*, 1–13.
23. Kyozuka, Y. An experimental study on the Darrieus-Savonius turbine for the tidal current power generation. *J. Fluid Sci. Technol.* **2008**, *3*, 439–449. [CrossRef]
24. Tescione, G.; Ragni, D.; He, C.; Ferreira, C.S.; Van Bussel, G. Near wake flow analysis of a vertical axis wind turbine by stereoscopic particle image velocimetry. *Renew. Energy* **2014**, *70*, 47–61. [CrossRef]
25. Balduzzi, F.; Drofelnik, J.; Bianchini, A.; Ferrara, G.; Ferrari, L.; Campobasso, M.S. Darrieus wind turbine blade unsteady aerodynamics: a three-dimensional Navier-Stokes CFD assessment. *Energy* **2017**, *128*, 550–563. [CrossRef]
26. Ferreira, C.S.; Van Kuik, G.; Van Bussel, G.; Scarano, F. Visualization by PIV of dynamic stall on a vertical axis wind turbine. *Exp. Fluids* **2009**, *46*, 97–108. [CrossRef]
27. Balduzzi, F.; Bianchini, A.; Maleci, R.; Ferrara, G.; Ferrari, L. Blade design criteria to compensate the flow curvature effects in H-Darrieus wind turbines. *J. Turbomach.* **2015**, *137*, 011006. [CrossRef]
28. Rezaeiha, A.; Kalkman, I.; Blocken, B. Effect of pitch angle on power performance and aerodynamics of a vertical axis wind turbine. *Appl. Energy* **2017**, *197*, 132–150. [CrossRef]
29. Kumar, P.M.; Surya, M.M.R.; Srikanth, N. Comparitive CFD analysis of darrieus wind turbine with NTU-20-V and NACA0018 airfoils. In Proceedings of the 2017 IEEE International Conference on Smart Grid and Smart Cities (ICSGSC), Singapore, 23–26 July 2017; IEEE: Piscataway, NJ, USA, 2017; pp. 108–114.
30. Rezaeiha, A.; Kalkman, I.; Montazeri, H.; Blocken, B. Effect of the shaft on the aerodynamic performance of urban vertical axis wind turbines. *Energy Convers. Manag.* **2017**, *149*, 616–630. [CrossRef]

31. Fujisawa, N.; Gotoh, F. Visualization study of the flow in and around a Savonius rotor. *Exp. Fluids* **1992**, *12*, 407–412. [CrossRef]
32. Kumar, P.M.; Rashmitha, S.R.; Srikanth, N.; Lim, T.-C. Wind Tunnel Validation of Double Multiple Streamtube Model for Vertical Axis Wind Turbine. *Smart Grid Renew.Energy* **2017**, *8*, 412–424. [CrossRef]
33. Kumar, P.M.; Ajit, K.R.; Srikanth, N.; Lim, T.-C. On the Mathematical Modelling of Adaptive Darrieus Wind Turbine. *J. Power Energy Eng.* **2017**, *5*, 133–158. [CrossRef]
34. Kumar, P.M.; Anbazhagan, S.; Srikanth, N.; Lim, T.-C. Optimization, Design, and Construction of Field Test Prototypes of Adaptive Hybrid Darrieus Turbine. *J. Fundam. Renew. Energy Appl.* **2017**, *7*, 245.
35. Rezaeiha, A.; Kalkman, I.; Blocken, B. CFD simulation of a vertical axis wind turbine operating at a moderate tip speed ratio: guidelines for minimum domain size and azimuthal increment. *Renew. Energy* **2017**, *107*, 373–385. [CrossRef]

© 2019 by the authors. Licensee MDPI, Basel, Switzerland. This article is an open access article distributed under the terms and conditions of the Creative Commons Attribution (CC BY) license (http://creativecommons.org/licenses/by/4.0/).

Article

Numerical Investigation of Air-Side Heat Transfer and Pressure Drop Characteristics of a New Triangular Finned Microchannel Evaporator with Water Drainage Slits

Brice Rogie *, Wiebke Brix Markussen, Jens Honore Walther and Martin Ryhl Kærn

Department of Mechanical Engineering, Technical University of Denmark, Nils Koppels Allé, Building 403, 2800 Kongens Lyngby, Denmark; wb@mek.dtu.dk (W.B.M.); jhw@mek.dtu.dk (J.H.W.); pmak@mek.dtu.dk (M.R.K.)
* Correspondence: brogie@mek.dtu.dk; Tel.: +45-45-25-41-21

Received: 14 October 2019; Accepted: 4 December 2019; Published: 11 December 2019

Abstract: The present study investigated a new microchannel profile design encompassing condensate drainage slits for improved moisture removal with use of triangular shaped plain fins. Heat transfer and pressure drop correlations were developed using computational fluid dynamics (CFD) and defined in terms of Colburn j-factor and Fanning f-factor. The microchannels were square 2.00 × 2.00 mm and placed with 4.50 mm longitudinal tube pitch. The transverse tube pitch and the triangular fin pitch were varied from 9.00 to 21.00 mm and 2.50 to 10.00 mm, respectively. Frontal velocity ranged from 1.47 to 4.40 m·s^{-1}. The chosen evaporator geometry corresponds to evaporators for industrial refrigeration systems with long frosting periods. Furthermore, the CFD simulations covered the complete thermal entrance and developed regions, and made it possible to extract virtually infinite longitudinal heat transfer and pressure drop characteristics. The developed Colburn j-factor and Fanning f-factor correlations are able to predict the numerical results with 3.41% and 3.95% deviation, respectively.

Keywords: microchannel; evaporator; water drainage; heat transfer; pressure drop; CFD

1. Introduction

Microchannel heat exchangers are attractive due to their high ratio of heat transfer area to internal volume. In recent years, they have gained increased market shares in many refrigeration and air-conditioning applications as air-cooled condensers, because of better thermo-hydraulic performance and compactness compared with traditional finned tube heat exchangers. However, their use as evaporators in refrigeration systems is challenged by (1) water condensate retention and (2) poor refrigerant distribution. The first point is extremely important in frosting conditions, since any retained water after a defrost cycle will simply freeze again on the evaporator surface.

A recent development by SAPA (now Hydro) Precision Tubing called Web-MPE offers a compromise between compactness and condensate retention, claiming a reduction of 90% water retained compared with traditional microchannel design with louvered fins [1]. The new microchannel profile designs are made with specialized drain paths in between each microchannel port, which means that the coil becomes thicker in the airflow direction.

The aim of the current work is to provide airside heat transfer and pressure drop correlations that are applicable for the design of novel ammonia microchannel evaporators for industrial refrigeration systems, e.g., cold stores, blast freezers etc., where the evaporator operates in freezing conditions. Such ammonia evaporators are traditionally finned-tube evaporators and employ large tube diameters, large tube pitches, and large fin pitches resulting in large frosting periods (up to 24 h). The air velocity and the air throw length are high in these evaporators, hence, the tube circuitry is commonly inline to provide a low airside pressure drop. Fin types are typically limited to plain fins or wavy fins since other fin types (louver fins, offset fins) result in higher pressure drop and/or ice formation in the opening sections of the fins thus reducing their significance.

Charge minimization in ammonia refrigeration systems is pertinent due to safety restrictions associated to these systems. National authorities have implemented regulations to restrict the amount of charge in industrial refrigeration systems in many countries. Today the charge limit in Denmark is 5000 kg. Exceeding this limit leads to significant increase in cost of the plant, and installation, maintenance, and operation costs, due to increased safety precautions. It provides an incentive for academics, refrigeration engineers, and equipment manufacturers to target their research and development towards low-charge ammonia equipment, including the evaporator.

In the current paper, the backbone (airside heat transfer and pressure drop correlations) of a completely new type of low charge ammonia evaporator is developed by means of Computational Fluid Dynamics (CFD). The work is based on vertically oriented Web-MPE profiles with use of triangular plain fins. Compared to other fin geometries, such as plain fins or wavy fins, the triangular plain fins allow water to drain due to their vertical inclination. They result in less pressure drop compared with wavy fins, offset, and louvered fins. The correlations developed herein may be used by refrigeration engineers and researchers to design and optimize novel ultra-low charge ammonia evaporators. To the author's best knowledge, no previous investigations exist in the open literature considering the thermo-hydraulic characteristics of this novel microchannel design.

Computational Fluid Dynamics (CFD) has become a major tool in order to investigate the flow behavior and/or thermo-hydraulic performance inside compact heat exchangers having various fin types such as louvered fins [2–5], offset fins [6–8], wavy fins [9–11], helically wound finned-tube bundles [12,13], and plain fins [14,15]. The results of CFD simulations can be used to correlate the thermo-hydraulic performances, generally defined in terms of the Colburn j-factor and the Fanning f-factor. Chennu and Paturu [16] performed CFD simulations in order to develop air-side correlations for offset fins. They developed their correlations distinctively for laminar and turbulent regions. Ismail and Velraj [10] undertook similar work considering offset fins and wavy fins. Bacellar et al. [17] used CFD simulations to develop air-side correlations of a compact finned tube heat exchanger with staggered tube arrangement without fins. Damavandi et al. [11] expressed the air-side characteristics of a wavy fin-and-elliptical tube heat exchanger. They used neural network to express the j- and f-factors with the aim to optimize the geometry with using a j vs. f Pareto front. Deng [18] conducted CFD simulations using Large Eddy Simulations (LES) to improve correlations for flat tubes and louvered fins. Similarly, Sadeghianjahromi et al. [19] developed correlations for a finned tube heat exchanger with louvered fins, focusing on the effect of louver angle. The above references employ the effectiveness-NTU method or LMTD method with mass flow averaged temperatures to extract the j-factors. These methods incorporate the hydraulic and thermal entrance region. Other researchers assume fully developed flow and use stream-wise periodic boundary conditions, first proposed by Patankar et al. [20], for simplifying the computational domain. For example, Martinez-Espinosa et al. [21] made fully developed flow correlations for compact finned-tube heat exchangers having helically segmented finned tubes. Recent reviews on the performances of various compact heat exchanger designs can be found in Awais and Bhuiyan [22] and Qasem and Zubair [23], considering various fin types and both experimental and numerical data.

The present study investigates the new microchannel evaporator design. The objective is to establish heat transfer and pressure drop correlations in terms of Colburn j-factor and Fanning f-factor, for use in two-stream compact heat exchanger simulation and optimization codes. The correlations do not consider frosting or defrosting conditions, even though the microchannel profile has been developed herein to solve the problem of water condensate retention in evaporators during defrost. The aim of the work is rather to provide the scientific foundation that allows engineers and researchers to design prototypes to be tested experimentally in frosting and defrosting conditions. Thus, the heat transfer and pressure drop in these conditions are subject for future work.

The paper is organized as follows: Section 2 describes the microchannel geometry, the CFD simulation design, modeling setup and verification, as well as data reduction methodology. Section 3 reports the results in terms of the correlations developed. In Section 4, the results and methodology are discussed. Finally, this is followed up by the conclusions in Section 5.

2. Method

2.1. Geometry of the Microchannel Evaporator

The microchannel evaporator is illustrated in Figure 1. It employs internal upward two-phase evaporating flow and external horizontal air crossflow. The triangular fins and drainage slits lead the water condensate downward through the evaporator during defrost.

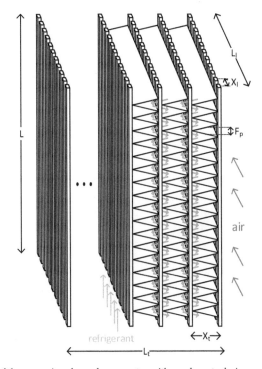

Figure 1. Sketch of the new microchannel evaporator with condensate drainage paths (blue arrows).

The microchannels were extruded and punched aluminum profiles. The extruded profiles had fixed inner and outer tube dimensions and fixed longitudinal tube pitch (Figure 2a) corresponding to the extrusion counterpart. After the extrusion process, the profiles were punched to remove a large part of the tube fins bridging the tubes in order to accommodate water drainage during defrost or dehumidifying conditions (Figure 2b). The remaining tube fins bridging the tubes were assumed to have a negligible contribution to the airside heat transfer, and thus excluded in the CFD simulations. Furthermore, the triangular plain fins had a fin thickness of 0.1625 mm.

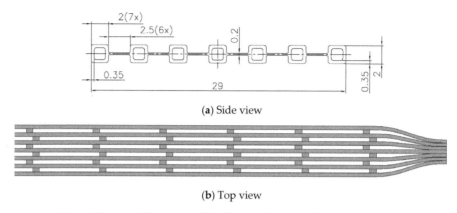

(a) Side view

(b) Top view

Figure 2. Extruded aluminum profile before punching (**a**) and after punching (**b**).

2.2. CFD Simulation Points

The work was based on the microchannel profile in Figure 2. This meant that the tube height/width and longitudinal pitch were fixed in the current work. With these parameters fixed, it was only the transverse tube pitch (X_t), the longitudinal length (L_l) (or the number of tube rows) and the fin pitch (F_p) that influenced the air-side heat transfer and pressure drop. A 3D model of the microchannel heat exchanger is shown in Figure 3. The fin angle (φ) was further dictated by the transverse tube pitch and fin pitch, respectively.

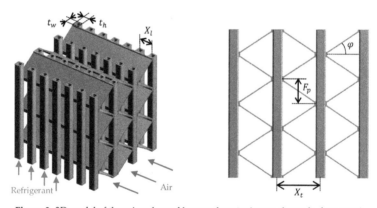

Figure 3. 3D model of the microchannel heat exchanger (seven channels, four rows).

For large fin pitches, considered in the current investigation to accommodate long frosting periods, the entrance region was found to be significant and therefore it was ensured to simulate enough longitudinal tubes (or tube rows) to establish fully developed hydraulic and thermal flow. Following this approach and to reduce the number of CFD simulations, the tube local friction and heat transfer coefficients were extracted in order to extent the global friction and heat transfer coefficients to even larger longitudinal lengths (see Section 2.4 (data reduction)). The parameterized geometry and frontal air velocity may be observed in Table 1. Some combinations of geometrical parameters were omitted (X_t = 9 mm, F_p = 7.5 mm) and (X_t = 9 mm, F_p = 10 mm) to avoid fin bending in the assembling and soldering process. The criteria used was fin angles less than 45°. In total, 42 simulations were carried out.

Table 1. Heat exchanger parameterization.

Parameters	Variation
Transverse tube pitch, X_t (mm)	9.00, 13.00, 17.00, 21.00
Fin pitch, F_p (mm)	2.50, 5.00, 7.50, 10.00
Longitudinal tube pitch, X_l (mm)	4.50
Tube rows, N_l (-)	35
Tube height, t_h (mm)	2.00
Tube width, t_w (mm)	2.00
Fin thickness, F_t (mm)	0.1625
Frontal air velocity, U_{fr} (m·s^{-1})	1.47, 2.93, 4.40

The hydraulic diameter (d_h) and the compactness ($\beta = 4\sigma/A_{tot}$) of the microchannel geometries were compared with a baseline plain finned-tube industrial refrigeration evaporator in Figure 4a,b. The baseline is outlined in Kristófersson et al. [24,25]. The tube diameter was 15.6 mm, the tube layout was inline 50 × 50 mm, the fin thickness was 0.35 mm and the fin pitch was 12 mm, and varied from 12 to 2.5 mm to represent a comparison at a similar fin pitch.

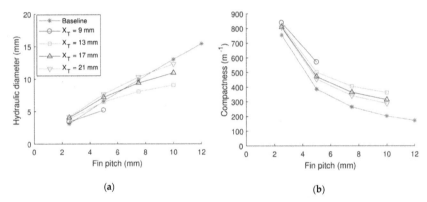

Figure 4. Hydraulic diameter (**a**) and compactness (**b**) vs. fin pitch of the microchannel evaporator geometries (Table 1) compared with the baseline finned-tube industrial refrigeration evaporator.

The hydraulic diameter of the microchannel geometries is nearly the same as the baseline finned-tube evaporator, while having more decreasing inclination as function of the fin pitch. Smaller fin pitch and transverse tube pitch result in smaller hydraulic diameters. The compactness is greater for the microchannel geometries, especially at higher fin pitch and lower transverse tube pitch compared with the baseline finned-tube evaporator. The greater differences between the hydraulic diameter and the compactness are due to the area contraction ratio (σ), which is smaller for the baseline finned-tube evaporator.

2.3. CFD Modeling Setup

CFD simulation brings an extensive knowledge of the flow behavior inside the microchannel evaporator and provides local data, which is challenging, and possibly subject to high uncertainties with an experimental setup. The CFD simulations were carried out using the commercial software ANSYS 19.1 with the CFX solver.

2.3.1. Modeling

In order to keep a reasonable simulation time, only a small part of the microchannel was modelled in the CFD simulations. Symmetries were used where the geometry allowed for it. The 3D CFD model is shown in the Figure 5. Moreover, only a single fin was included in the computational domain.

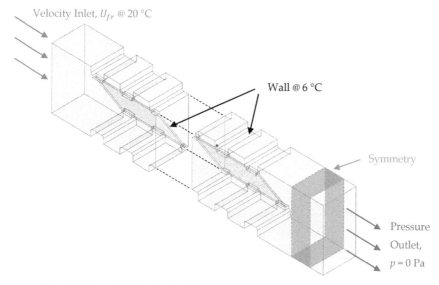

Figure 5. 3D model of the simulated geometry (X_t = 9.00 mm, F_p = 5.00 mm, N_l = 35).

The tube and fin walls were assumed to have a constant wall temperature (6 °C) consistent with the use of the effectiveness-NTU method for single stream heat exchangers, which was used to calculate the tube local and global heat transfer coefficients. Moreover, the temperature values of the wall and air inlet are independent on the heat exchanger effectiveness, which is valid as long as the air properties can be assumed constant. The derived heat transfer coefficients were, therefore, tube and fin surface averaged. The constant wall temperature means that the heat conduction through the metal (tubes and fins) was disregarded in the CFD calculations, and that it must be included when using the heat transfer correlations. In Appendix A.1, it is demonstrated that the fin efficiency for rectangular fins can be used to model the heat conduction with good accuracy, even though a heat flux concentration (2D effect) occurs near the base of the fin at the microchannel walls. Furthermore, the no-slip condition was employed at the walls, and symmetry condition at the four lateral surfaces. The air was assumed an incompressible ideal gas due to the small temperature changes.

2.3.2. Mesh Analysis

For flow around obstacles, the laminar boundary layer restarts at the tip of each tube, with a transitional flow in their wakes due to vortex formation. For inline rectangular tube configuration, the heat transfer rate is expected to be highest at the leading corner edge of each tube while decreasing along the tube longitudinally. In the wake region, recirculation zones typically appear with lower velocities and heat transfer rate. However, turbulent vortices improve the mixing and increase the heat transfer in the neighborhood regions too [26].

The restart of the boundary layer principle is similar for offset fins, which generally provide a very good heat transfer rate compared to other fin designs [23]. The transition from laminar to turbulent flow may appear for low Reynolds number, $Re < 500$, such as described by Sahiti et al. [27]. The range of Reynolds numbers in the current simulations is from 500 to 4000, therefore the k-ω SST turbulence model, based on the work of Menter [28], was selected. Kim et al. [29] showed that the k-ω SST turbulence model gives better performances, compared to the k-ε and realizable k-ε turbulence models, in terms of predicted j and f factors for offset fins at $Re > 1000$. Finally, Chimres et al. [30] showed that the k-ω SST turbulence model results in good agreement with experimental heat transfer and pressure drop data for flow around tubes.

A mesh sensitivity analysis was performed. The y^+ was kept below one to ensure accurate resolution of the viscous boundary layer, advised by the ANSYS user guide [31] when using the k-ω SST turbulence model. The size of the computational grid was analyzed in order to ensure the grid independence. The values of the global Colburn j-factor and the Fanning f-factor are shown in Figure 6 as function of the mesh size.

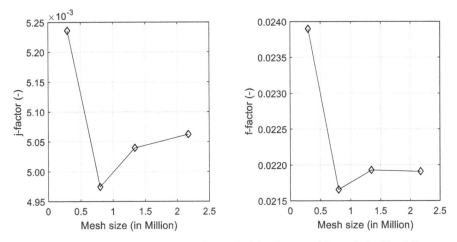

Figure 6. Colburn j-factor (**left**) and friction f-factor (**right**) as function of the mesh size (X_t = 9.00 mm, F_p = 5.00 mm, U_{fr} = 4.40 m·s^{-1}).

The difference between two consecutive values of the Colburn j-factor and the Fanning f-factor is lower than 0.5% from 1.3 to 2.2 M elements. Therefore, the mesh of 1.3 M elements was selected to have a good balance between accuracy and calculation speed. The 1.3 M mesh is shown in Figure 7.

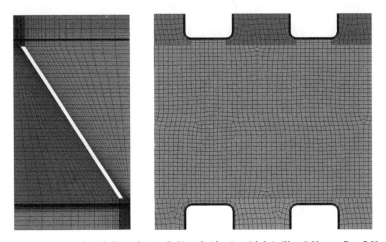

Figure 7. Computational grid. Frontal view (**left**) and side view (**right**), (X_t = 9.00 mm; F_p = 5.00 mm).

The mesh was fully structured (only hexahedral elements) to minimize numerical diffusion. Furthermore, the mesh was refined close to the wall to keep the $y^+ < 1$.

2.3.3. Velocity, Temperature, and Pressure Profiles

The longitudinal velocity, the temperature, and the static pressure contours are shown in Figures 8 and 9, respectively, for the simulation: $X_t = 9.00$ mm, $F_p = 5.00$ mm, $U_{fr} = 4.40$ m·s^{-1}. Figure 8 shows the contours at different locations of the heat exchanger, i.e., the two first tubes (entrance region), the 17th and 18th tubes (center) and the last two tubes (exit region), respectively. Figure 9 shows the contours at different minimum cross sections normal to the airflow, i.e., the first tube (entrance region), the 17th tube (center), and the last tube (exit region), respectively.

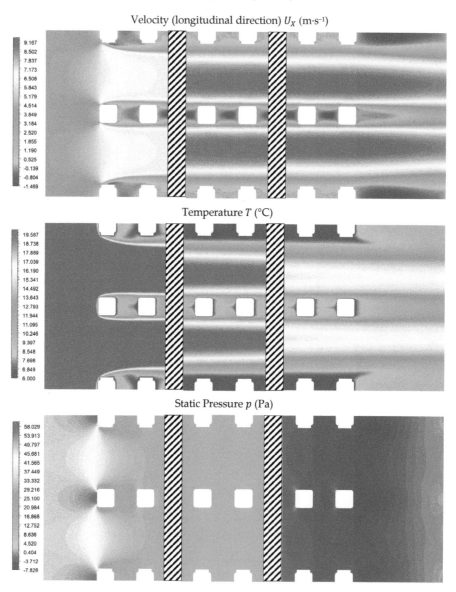

Figure 8. Longitudinal velocity, temperature, and static pressure contours at the entrance (**left**), middle (**center**), and exit (**right**); (side view); $X_t = 9.00$ mm, $F_p = 5.00$ mm, $U_{fr} = 4.40$ m·s^{-1}.

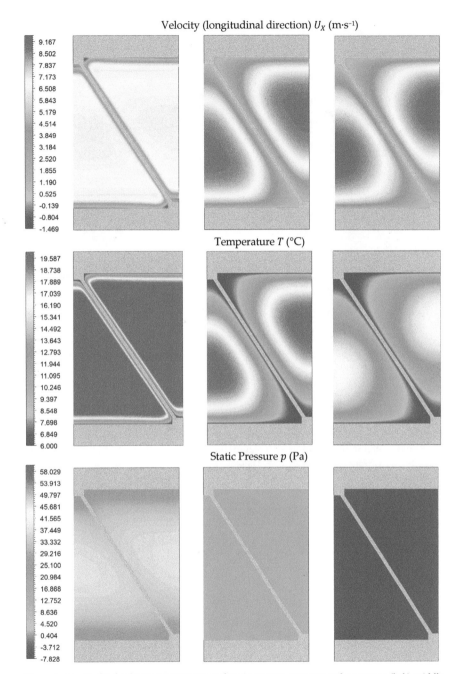

Figure 9. Longitudinal velocity, temperature, and static pressure contours at the entrance (**left**), middle (**center**), and exit (**right**); (frontal view); $X_t = 9.00$ mm, $F_p = 5.00$ mm, $U_{fr} = 4.40$ m·s^{-1}.

The velocity contours on Figures 8 and 9 indicate that the flow develops and reaches almost fully developed velocity contour at the center region compared with the exit region. The temperature contours indicate similarity at the center and the exit region, which also confirm that the flow becomes fully developed. Additionally, the pressure change during the inlet contraction and outlet expansion are easily observable in Figure 8. Figure 10 indicates recirculation in the wake of the channels with locally low heat transfer coefficient.

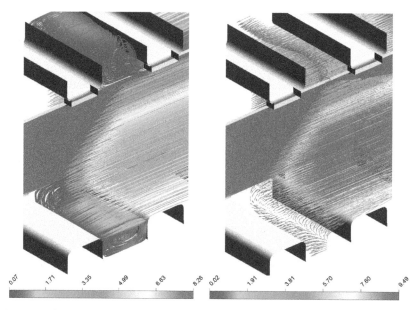

Figure 10. Velocity streamlines (**left**) and velocity vectors (**right**); X_t = 9.00 mm, F_p = 5.00 mm, U_{fr} = 4.40 m·s^{-1}.

2.4. Data Reduction

The data reduction followed simple equations to calculate the involved surface areas and flow areas etc., and matched the CFD implementation including rounding effects within 2% deviation (including the maximum core velocity, U_c). These equations, considering the computational domain, are given as follows:

$$A_{fr} = X_t \cdot F_p, \tag{1}$$

$$A_c = (X_t - t_h) \cdot F_p - \frac{P_f \cdot F_t}{2}, \tag{2}$$

$$P_f = 2 \cdot \left(\left[(X_t - t_h)^2 + F_p^2 \right]^{1/2} - F_t \right), \tag{3}$$

$$A_f = P_f \cdot L_l, \tag{4}$$

$$A_{tube} = \left[2 \cdot (t_w + t_h) \cdot F_p - 2 \cdot F_t \cdot t_w \right] \cdot N_l, \tag{5}$$

$$A_{tot} = A_{tube} + A_f, \tag{6}$$

where A_{fr} is the frontal area, A_c the minimum free flow area, P_f the fin perimeter, A_f the fin area, A_{tube} the bare tube area, A_{tot} the total heat transfer area. To calculate the Colburn j-factor, the effectiveness-NTU method was used with the assumption of constant wall temperature,

$$NTU_{air} = -ln\left(1 - \frac{T_o - T_i}{T_w - T_i}\right), \tag{7}$$

$$h = NTU_{air} \cdot \frac{C_{min}}{A_{tot}}, \tag{8}$$

$$j = h \cdot \frac{Pr^{\frac{2}{3}}}{\rho \cdot U_c \cdot c_p}, \tag{9}$$

where T_o, T_i, and T_w are the outlet, inlet, and wall temperature, respectively, h the heat transfer coefficient, C_{min} the minimum heat capacitance rate, Pr the Prandlt number, ρ the density, and c_p the specific heat capacity at constant pressure.

These equations were used to calculate the tube local heat transfer coefficient and global heat transfer coefficients, respectively. The tube local heat transfer coefficients were based on the mass-flow averaged inlet and outlet air temperatures of each tube row and local surface area. On the other hand, the global heat transfer coefficient was based on the mass flow averaged inlet temperature of the heat exchanger and the mass flow averaged outlet temperature of each tube row and cumulated local area.

Figure 11 illustrates the tube local and global heat transfer coefficients, and the extended global heat transfer coefficient, calculated by further integrating the fully developed tube local heat transfer coefficient:

$$h_{ext} = \frac{1}{L_{l,sim}} \cdot \int_0^{L_{l,sim}} h_{loc} \cdot dL_l + \frac{1}{L_l - L_{l,sim}} \cdot \int_{L_{l,sim}}^{L_l} h_{fd,loc} \cdot dL_l, \tag{10}$$

where $L_{l,sim}$ is the longitudinal length of the simulated geometry, h_{loc} and $h_{fd,loc}$ are the local and fully developed local heat transfer coefficient, respectively. The extended global heat transfer coefficient was integrated to provide global heat transfer coefficients for 90 tube rows in total for each of the 42 CFD simulations.

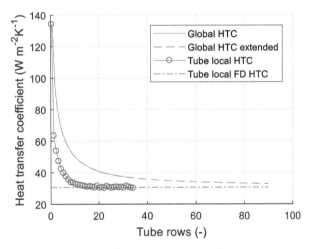

Figure 11. Extension of the global heat transfer coefficient (X_t = 13.00 mm, F_p = 7.50 mm, U_{fr} = 4.40 m·s^{-1}).

The standard deviation of the five rearmost tube rows of the simulated geometry in terms of tube local heat transfer coefficients range from 0.07 to 1.23 W·m^{-2}·K^{-1} for all the considered simulations. These values were considered reasonable for assuming thermally developed flow.

Similarly, the total pressure drop was reconstructed and extended by calculating the contraction and expansion pressure drop at the inlet (i) and the outlet (o), as well as the local core pressure drop (core),

$$\Delta p_{tot} = \Delta p_i + \Delta p_{core} + \Delta p_o, \tag{11}$$

$$\Delta p_i = \frac{G_c^2}{2 \cdot \rho_i} \cdot \left(1 - \sigma^2 + K_c\right), \tag{12}$$

$$\Delta p_o = \frac{G_c^2}{2\cdot\rho_i}\left(1-\sigma^2-K_e\right)\cdot\frac{\rho_i}{\rho_o}, \tag{13}$$

$$\Delta p_{core} = \frac{G_c^2}{2\cdot\rho_i}\left[f\cdot\frac{A_{tot}}{A_c}\cdot\frac{\rho_i}{\rho_m}+2\cdot\left(\frac{\rho_i}{\rho_o}-1\right)\right], \tag{14}$$

where Δp is the pressure drop, G_c the maximum mass velocity, σ the contraction ratio, K_c and K_e the contraction and expansion coefficient, respectively, and ρ_m the mean density. Here it was assumed that the contraction and expansion pressure drops were independent on the number of tube rows and could be directly added to the averaged local core pressure. Furthermore, the acceleration pressure drop in Equation (14) was assumed negligible. Figure 12 illustrates the simulated pressure drop, the reconstruction, and extension of the reconstruction.

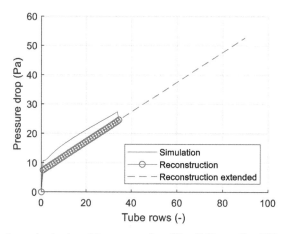

Figure 12. Reconstruction and extension of the pressure drop (X_t = 13.00 mm, F_p = 7.50 mm, U_{fr} = 4.40 m·s^{-1}).

Finally, the extended pressure drop was converted into a total friction factor, which incorporates the contraction and the expansion pressure drops, respectively, consistent with usual practice regarding compact heat exchanger pressure drop correlations. This was done by solving Equations (11)–(14) for f with $K_c = K_e = 0$ and negligible core acceleration pressure drop (term 2 in Equation (14)).

Appendix A.2 demonstrates that the extension of the global heat transfer coefficient, as well as the reconstruction and extension of the pressure drop, are indeed valid by simulating geometrical designs with 18, 35, 53, and 70 tube rows. Moreover, the results were almost identical and independent of the number of tube rows.

3. Results

3.1. Heat Transfer and Pressure Drop Regression

The reduced CFD results in terms Colburn j-factor and Fanning friction f-factor were regressed using multiple linear and nonlinear regression techniques. Moreover, the asymptotic model was used to model the transition between the entrance region (*ent*) and the fully developed (fd) region, respectively,

$$y^n = y_{ent}^n + y_{fd}^n, \tag{15}$$

where y denote the Colburn j-factor or Fanning f-factor, respectively. Four nondimensional parameters based on the hydraulic diameter were used to model the entrance and fully developed regions,

$$y_{ent} = b_1\cdot Re_{d_h}^{b2}\cdot\left(\frac{L_l}{d_h}\right)^{b3}\cdot\left(\frac{X_t}{d_h}\right)^{b4}\cdot\left(\frac{F_p}{d_h}\right)^{b5}, \tag{16}$$

$$y_{fd} = b_6 \cdot Re_{d_h}^{b_7} \cdot \left(\frac{X_t}{d_h}\right)^{b_8} \cdot \left(\frac{F_p}{d_h}\right)^{b_9}, \quad (17)$$

where Re is the Reynolds number, $b_{1,2...}$ regression coefficients, and d_h the hydraulic diameter given by,

$$d_h = \frac{4 \cdot A_c \cdot L_l}{A_{tot}}. \quad (18)$$

Notice that the fully developed equation was independent longitudinally, in contrast to the entrance equation. The regression procedure followed the four steps:

1. Linear regression of y_{ent} based on the first five consecutive points longitudinally (the choice of five points was based on visual interpretation of the results),
2. Linear regression of y_{fd},
3. Nonlinear regression of y,
4. A repeated nonlinear regression of the coefficients b_1, b_2, b_3, b_4, b_5 and n in order to alleviate errors related to the visual interpretation in step 1.

Figure 13 illustrates the regression methodology for the j- and f-factor, respectively.

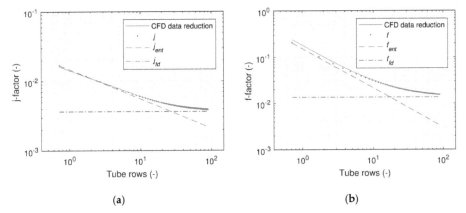

Figure 13. Regression methodology for j-factor (**a**) and friction factor (**b**). (X_t = 13.00 mm, F_p = 7.50 mm, U_{fr} = 4.40 m·s^{-1}).

Equations (15)–(17) resulted in very accurate correlations compared with the CFD simulation results. Table 2 indicates the coefficients to be used for the j- and f-factor correlations and Figure 14 shows the resulting parity plots. A total number of 42 × 90 = 3780 simulations points were used to derive the heat transfer and pressure drop correlations.

Table 2. Coefficients for the heat transfer and friction correlations.

Coefficient	j-Factor	f-Factor
b_1	0.8539	0.8665
b_2	−0.5433	−0.2804
b_3	−0.4234	−0.8512
b_4	0.0424	0.1777
b_5	−0.0966	0.9961
b_6	0.0303	1.4393
b_7	−0.2697	−0.5795
b_8	0.1015	−0.1196
b_9	0.1095	−0.2454
n	3.1784	1.2611

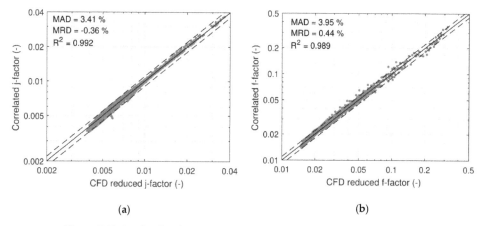

Figure 14. Parity plots for j-factor (**a**) and f-factor (**b**), dashed lines indicate 10% error.

Furthermore, the parity plots indicate the mean average deviation (*MAD*), mean relative deviation (*MRD*), and coefficient of determination (R^2). These values were computed by using the following equations:

$$MAD = \frac{1}{n} \cdot \sum_{i=1}^{n} \left| \frac{y_{i,pred} - y_{i,sim}}{y_{i,sim}} \right|, \tag{19}$$

$$MRD = \frac{1}{n} \cdot \sum_{i=1}^{n} \left(\frac{y_{i,pred} - y_{i,sim}}{y_{i,sim}} \right), \tag{20}$$

$$R^2 = 1 - \frac{\sum_{i=1}^{n}\left(y_{i,sim} - y_{i,pred}\right)^2}{\sum_{i=1}^{n}\left(y_{i,sim} - \overline{y}_{sim}\right)^2}, \tag{21}$$

where n is the number of samples, and *pred* and *sim* denote prediction and simulation, respectively. The accuracy of the correlations cannot be guaranteed when the correlations are applied beyond the ranges of the simulation points. The ranges of the simulation points were as follows:

- $d_h = 3.45 \text{ mm} - 12.33 \text{ mm}$
- $Re_{d_h} = 481 - 4084$
- $X_t/d_h = 1.4 - 5.0$
- $F_p/d_h = 0.6 - 1.1$

3.2. Analysis of Entrance Region

The results of this work indicated that the heat transfer effects of the entrance region are significant and necessary to include in the heat transfer correlation. The thermally developed region is typically claimed when the heat transfer coefficient is within 98% of the fully developed value. Figure 15 shows the number of tube rows for which this criterion is reached at different frontal velocities as a function of the hydraulic diameter, the Reynolds number, and the fin angle.

The results show that the thermally developed flow criterion is reached at different tube rows depending on mainly the air velocity and hydraulic diameter. The highest entrance regions are found at low air velocity and high hydraulic diameter and vice versa. No particular tendencies are found with respect to fin angle.

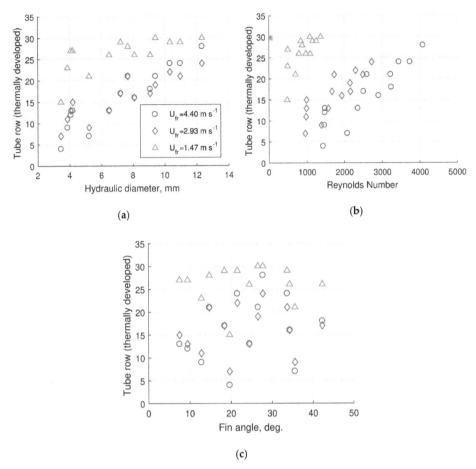

Figure 15. Number of tube rows to reach thermally developed flow at different frontal velocities vs. hydraulic diameter (**a**), Reynolds number (**b**), and fin angle (**c**).

3.3. Volume Goodness Factor

The volume goodness factor, defined for extended surfaces by Shah and Sekulic [32], is used to compare the microchannel geometries with the baseline finned-tube evaporator for industrial refrigeration (see Section 2.2 for comparisons of hydraulic diameter and compactness). The volume goodness factor compares the heat transfer rate per unit temperature difference and unit core volume versus the friction power expenditure per unit core volume, both defined by:

$$\eta_o \cdot h \cdot \beta = \frac{c_p \cdot \mu}{Pr^{2/3}} \cdot \eta_o \cdot \frac{4 \cdot \sigma}{d_h^2} \cdot j \cdot Re, \tag{22}$$

$$E \cdot \beta = \frac{\mu^3}{2 \cdot \rho^2} \frac{4 \cdot \sigma}{d_h^4} \cdot f \cdot Re^3, \tag{23}$$

where η_o is the overall surface efficiency calculated using the fin efficiency for rectangular fins (see Appendix A.1), μ is the viscosity, and E is the friction power per unit surface area.

Most correlations for finned-tube evaporators in the literature are developed for staggered tube layouts as pointed out by Webb and Kim [33]. The correlations are typically developed for designs with lower fin pitch and lower number of tube rows compared with the baseline finned-tube evaporator for industrial refrigeration. This complicates the choice of correlations to compare with our results. In the following comparison, the plain finned-tube correlations by Kaminski and Groß [34] are used to calculate the j- and f- factors and the overall surface efficiency, as outlined by Fraß et al. [35]. Figure 16 shows the comparisons of the microchannel evaporator having 35 tube rows and the baseline finned-tube evaporator having eight tube rows.

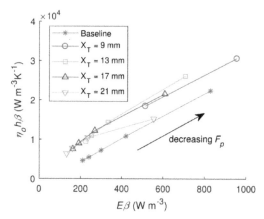

Figure 16. Volume goodness factors for the microchannel evaporator geometries (Table 1) and the baseline finned-tube industrial refrigeration evaporator (U_{fr} = 2.93 m·s^{-1}, N_l = 35 and 8, respectively).

The volume goodness factors reveal that the microchannel evaporator is indeed more attractive than the baseline finned-tube evaporator, transferring more heat per unit volume at the same fluid flow power, and vice versa. In other words, the microchannel performs the best from the viewpoint of heat exchanger volume. There is however a single point (X_t = 21 mm, F_p = 2.5 mm) where the pressure drop of the microchannel evaporator increases more than the heat transfer, and results in similar performance as the baseline finned tube evaporator. This is mainly due to the low fin angle effects for this geometry. Furthermore, the variation of the number of tube rows had an insignificant effect on the volume goodness factor.

4. Discussion

The correlations obtained herein are based on (or fixed by) the microchannel profile design. For providing general correlations, the tube width, the tube height, and the longitudinal tube pitch must be parametrized too. This work did not attempt to reach beyond the actual dimensions of the extruded microchannel profile. The work must rather be viewed as a first attempt to deliver correlations for the design of such evaporators, and to be used for future research and development, especially devoted to the refrigerant charge minimization in industrial refrigeration systems. The developed correlations can be used to design the new microchannel evaporator for this purpose in dry conditions.

Frosting, defrosting, and water condensate drainage are furthermore dependent on the total size of the evaporator, especially the height as the water condensate need to travel downwards through the triangular fins. These considerations are considered for future work. A prototype evaporator is already outlined at this moment and it will be tested experimentally at the Danish Technological Institute laboratory in the near future. These tests will be used to compare the correlations accuracy. Furthermore, tests are planned to study the cooling capacity during frost build-up and defrost performances.

Additionally, the CFD simulations should be viewed as idealized flows compared with the total evaporator flow in a real installation. There are many peculiarities in real evaporators such as airside and tube-side temperature nonuniformity, fluid flow maldistribution, nonidealized fin conduction, transitional fluid flow regimes, imperfect contact between tubes and fins, fin geometry manufacturing uncertainties, etc. These factors must be incorporated in the anticipated uncertainty during the design of the microchannel evaporator.

The correlated heat transfer coefficient is surface averaged. To be used in heat exchanger simulation codes, it should be used to calculate the fin efficiency as well. In Appendix A.1, it is demonstrated that the fin efficiency for rectangular fins can be used with good accuracy, even though a heat flux concentration (2D effect) occurs near the base of the fin at the microchannel walls.

The entrance region was found to be significant in the current analysis. Disregarding the effect of the entrance region might lead to significant underestimations of the global heat transfer coefficient, especially at lower frontal velocities where the highest entrance regions were found. It should be stressed that the current investigation considers plain triangular fins with large fin pitches. The developing region might be insignificant for other types of fins and fin pitches, e.g., because of larger secondary flows in louvered fins. No clear entrance length trends were found in terms of Reynolds number or fin angle. However, Shah and London [36] found that the entrance region reached a minimum for triangular duct flow with angles around $2\varphi \approx 55°$.

Additionally, in Appendix A.2, the extension of the global heat transfer coefficient longitudinally as well as the reconstruction and extension of the pressure drop longitudinally are assessed and discussed. Indeed, the methodology can be applied to minimize CFD simulation points and simplify the computational domain.

5. Conclusions

This paper presented heat transfer and pressure drop correlations for a new microchannel evaporator design, based on a newly developed microchannel profile with condensate drainage slits and use of triangular shaped plain fins with large fin pitch. The chosen evaporator geometry corresponds to evaporators for industrial refrigeration systems with long frosting periods. Heat transfer and pressure drop correlations were developed using computational fluid dynamics (CFD) and defined in terms of Colburn j-factor and Fanning f-factor. The computational domain covered the complete thermal entrance and developed regions, which made it possible to extract virtually infinite longitudinal heat transfer and pressure drop characteristics. Indeed, the entrance region was found to be significant compared to the typical longitudinal evaporator length. Therefore, the asymptotic model was used to correlate the entrance and developed regions, respectively. The developed Colburn j-factor and Fanning f-factor correlations were able to predict the numerical results with 3.41% and 3.95% deviation, respectively.

Author Contributions: Writing—original draft, B.R. and M.R.K; Writing—review and editing, B.R., W.B.M., J.H.W. and M.R.K.

Funding: The research was funded by the Danish Energy Agency (DEA) through the Energy Technology Development and Demonstration Programme (EUDP), project number: 64017-05128—FARSevap.

Acknowledgments: The funding by the DEA is greatly acknowledged. Furthermore, the authors acknowledge the collaboration regarding the new microchannel evaporator design and fin design with our industrial partners: The Danish Technological Institute, Aluventa and Hydro Precision Tubing.

Conflicts of Interest: The authors declare no conflict of interest.

Nomenclature

A_c	Minimum free flow area, m^2
A_f	Fin Area, m^2
A_{fr}	Frontal area, m^2
A_{tot}	Total heat transfer area, m^2
$b_1, b_2 \ldots b_9$	Regression coefficients, (-)
c_p	Heat capacity at constant pressure, J kg^{-1}K^{-1}
C_{min}	Lowest heat capacity rate, W K^{-1}

d_h	Hydraulic diameter, m
E	Friction power per unit surface area W m^{-2}
f	Friction factor, (-)
F_p	Fin pitch, m
F_t	Fin thickness, m
G_c	Maximum mass velocity, kg s^{-1}m^{-2}
h	Heat transfer coefficient, W m^{-2}K^{-1}
j	Colburn factor, (-)
k_f	Thermal conductivity of fin, W m^{-1}K^{-1}
K_e, K_c	Expansion and Contraction coefficient, (-)
N_l	Tube rows, (-)
L_l	Longitudinal length, m
L_t	Transverse length, m
p	Pressure, Pa
P_f	Fin perimeter, m
Pr	Prandtl number, (-)
$\dot{Q}_{actual}$	Heat transfer to the fin, W
$\dot{Q}_{ideal}$	Heat transfer to an ideal fin, W
R^2	Coefficient of determination, (-)
Re	Reynolds number, (-)
t_h	Tube height, m
t_w	Tube width, m
T	Temperature, K
T_a	Fluid ambient temperature, K
T_b	Fin base temperature, K
T_f	Fin average temperature, K
U_c	Maximum air velocity, m s^{-1}
U_{fr}	Frontal air velocity, m s^{-1}
U_X	Flow velocity in x-direction, m s^{-1}
X_l	Longitudinal tube pitch, m
X_t	Transverse tube pitch, m
y^+	Dimensionless distance to the wall, (-)

Greek Symbols

β	Compactness, m^{-1}
η_f	Fin efficiency, (-)
η_o	Overall surface efficiency, (-)
μ	Viscosity, Pa s^{-1}
ρ	Fluid density, kg m^{-3}
σ	Contraction ratio, (-)
φ	Fin angle, deg
Δp	Pressure drop, Pa

Abbreviation

CFD	Computational Fluid Dynamics
FEM	Finite Element Method
LES	Large Eddy Simulation
LMTD	Logarithmic Mean Temperature Difference
MAD	Mean Absolute Deviation
MRD	Mean Relative Deviation
NTU	Number of Transfer Units
SST	Shear Stress Transport

Subscripts

ent	Entrance
fd	Fully developed
i	Inlet
m	Mean
o	Outlet
v	Vapor
w	Wall

Appendix A

Appendix A.1. Fin Efficiency

The computation of the heat transfer coefficient and fin efficiency are equally important for the design optimization of the new microchannel heat exchanger. In order to examine the fin efficiency, two heat conduction Finite Element (FEM) simulations were carried out, one with a smaller fin and another with a larger fin. Symmetry plans were used again to minimize the computational domain. The temperature of the channel internal walls was specified to 6 °C. A constant heat transfer coefficient was applied to the fin and channel external surfaces, corresponding to the thermally developed local heat transfer coefficient extracted from the CFD simulations. The contact between the channels and the fins is assumed perfect. The fin efficiency was calculated based on the results of the heat conduction simulations as follows.

$$\eta_f = \frac{\dot{Q}_{actual}}{\dot{Q}_{ideal}} = \frac{h \cdot \int_f (T_a - T_f) \cdot dA}{h \cdot A_f \cdot (T_a - T_b)}, \tag{A1}$$

where T_a is the mean fluid temperature, T_b is the fin base (or contact) temperature, and T_f is the fin temperature. The fin efficiency was compared with the analytical fin efficiency evaluated for rectangular fins

$$\eta_f = \frac{\tanh(m \cdot l_c)}{m \cdot l_c}, \tag{A2}$$

$$m = \left(\frac{2 \cdot h}{k_f \cdot F_t}\right)^{1/2}, \tag{A3}$$

with $l_c = P_f/4$. The comparison is shown in Figure A1 and the temperature contours of the heat conduction simulations are shown in Figure A2.

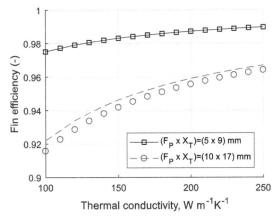

Figure A1. Fin efficiency vs. thermal conductivity for two geometries. Symbols indicate the analytical fin efficiency evaluated for rectangular fins.

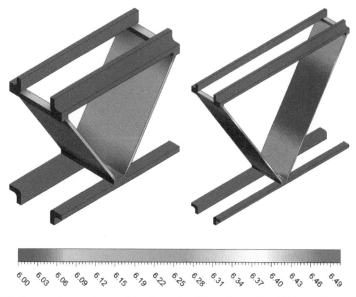

Figure A2. Temperature contours of the channel and fins. ($F_p \times X_t = 5.00 \times 9.00$ mm (left) and $F_p \times X_t = 10.00 \times 17.00$ mm (right), $U_{fr} = 4.40$ m·s^{-1}).

The results demonstrate that the analytical fin efficiency for rectangular fins can be used with good accuracy to model the fin efficiency of the current fin design. This holds true even though heat flux concentration (2D effects) occurs near the base of the fin at the microchannel walls.

Appendix A.2. Longitudinal Extrapolation Analysis

In this section, the validity of the extended global heat transfer coefficient is assessed. Moreover, three additional simulations were performed at different number of tube rows. The geometrically centered dimensions ($F_p = 7.50$ mm, $X_t = 13.00$ mm) and highest air velocity ($U_{fr} = 4.40$ m·s^{-1}) were used in these simulations. The results in terms of Colburn j-factor and the Fanning f-factor are represented in Figure A3 including the prediction of our correlation (Equation (15)).

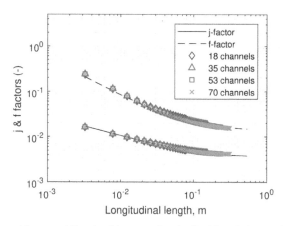

Figure A3. Colburn j-factor and Fanning f-factor vs. longitudinal length (or number of tube rows) ($F_p = 7.50$ mm, $X_t = 13.00$ mm, $U_{fr} = 4.40$ m·s^{-1}).

The results showed very good agreement between the results of simulations that were close to identical, and well predicted using the developed correlations. The MAE of the four simulated Colburn j-factors compared with Equation (15) were 1.9%, 1.9%, 2.4%, and 3.2% for the 18, 35, 53, and 70 tube rows, respectively. The MAE of the four simulated Fanning f-factors compared with Equation (15) were 8.2%, 3.7%, 2.2%, and 2.6% for the 18, 35, 53, and 70 tube rows, respectively. This indicated that 35 tube rows were sufficient for developing the correlations in the paper.

References

1. Ploug, O.; Vestergaard, B. New microchannel profile for evaporators and for refrigeration condensers. In Proceedings of the 4th International Congress on Aluminium Heat Exchanger Technologies for HVAC&R, Düsseldorf, Germany, 10–11 June 2015.
2. Atkinson, K.N.; Drakulic, R.; Heikal, M.R.; Cowell, T.A. Two- and three-dimensional numerical models of flow and heat transfer over louvred fin arrays in compact heat exchangers. *Int. J. Heat Mass Transf.* **1998**, *41*, 4063–4080. [CrossRef]
3. Perrotin, T.; Clodic, D. Thermal-hydraulic CFD study in louvered fin-and-flat-tube heat exchangers. *Int. J. Refrig.* **2004**, *27*, 422–432. [CrossRef]
4. Qian, Z.; Wang, Q.; Cheng, J.; Deng, J. Simulation investigation on inlet velocity profile and configuration parameters of louver fin. *Appl. Therm. Eng.* **2018**, *138*, 173–182. [CrossRef]
5. Sarpotdar, S.; Nasuta, D.; Aute, V.; Systems, O.; Road, V.M.; Park, C. CFD Based Comparison of Slit Fin and Louver Fin Performance for Small Diameter (3 mm to 5 mm) Heat Exchangers. *Int. Compress. Eng. Refrig. Air Cond.* **2016**, *1988*, 1–10.
6. Jiang, Q.; Zhuang, M.; Zhu, Z.; Shen, J. Thermal hydraulic characteristics of cryogenic offset-strip fin heat exchangers. *Appl. Therm. Eng.* **2019**, *150*, 88–98. [CrossRef]
7. Martinez, E.; Vicente, W.; Salinas-Vazquez, M.; Carvajal, I.; Alvarez, M. Numerical simulation of turbulent air flow on a single isolated finned tube module with periodic boundary conditions. *Int. J. Therm. Sci.* **2015**, *92*, 58–71. [CrossRef]
8. Peng, H.; Ling, X.; Li, J. Performance investigation of an innovative offset strip fin arrays in compact heat exchangers. *Energy Convers. Manag.* **2014**, *80*, 287–297. [CrossRef]
9. Bhuiyan, A.A.; Amin, M.R.; Naser, J.; Islam, A.K.M. Effects of geometric parameters for wavy finned-tube heat exchanger in turbulent flow: A CFD modeling. *Front. Heat Mass Transf.* **2015**, *6*, 1–11. [CrossRef]
10. Ismail, L.S.; Velraj, R. Studies on fanning friction (f) and colburn (j) factors of offset and wavy fins compact plate fin heat exchanger-a CFD approach. *Numer. Heat Transf. Part A Appl.* **2009**, *56*, 987–1005. [CrossRef]
11. Damavandi, M.D.; Forouzanmehr, M.; Safikhani, H. Modeling and Pareto based multi-objective optimization of wavy fin-and-elliptical tube heat exchangers using CFD and NSGA-II algorithm. *Appl. Therm. Eng.* **2017**, *111*, 325–339. [CrossRef]
12. Cléirigh, C.T.Ó.; Smith, W.J. Can CFD accurately predict the heat-transfer and pressure-drop performance of finned-tube bundles? *Appl. Therm. Eng.* **2014**, *73*, 681–690. [CrossRef]
13. Lindqvist, K.; Næss, E. A validated CFD model of plain and serrated fin-tube bundles. *Appl. Therm. Eng.* **2018**, *143*, 72–79. [CrossRef]
14. Kemerli, U.; Kahveci, K. Numerical Investigation of Air-Side Heat Transfer and Fluid Flow in a Microchannel Heat Exchanger. In Proceedings of the 2nd World Congress on Mechanical, Chemical, and Material Engineering (MCM'16), Budapest, Hungary, 22–23 August 2016.
15. Kumar, A.; Joshi, J.B.; Nayak, A.K.; Vijayan, P.K. 3D CFD simulations of air cooled condenser-III: Thermal-hydraulic characteristics and design optimization under forced convection conditions. *Int. J. Heat Mass Transf.* **2016**, *93*, 1227–1247. [CrossRef]
16. Chennu, R.; Paturu, P. Development of heat transfer coefficient and friction factor correlations for offset fins using CFD. *Int. J. Numer. Methods Heat Fluid Flow* **2011**, *21*, 935–951. [CrossRef]
17. Bacellar, D.; Aute, V.; Huang, Z.; Radermacher, R. Airside friction and heat transfer characteristics for staggered tube bundle in crossflow configuration with diameters from 0.5 mm to 2.0 mm. *Int. J. Heat Mass Transf.* **2016**, *98*, 448–454. [CrossRef]

18. Deng, J. Improved correlations of the thermal-hydraulic performance of large size multi-louvered fin arrays for condensers of high power electronic component cooling by numerical simulation. *Energy Convers. Manag.* **2017**, *153*, 504–514. [CrossRef]
19. Sadeghianjahromi, A.; Kheradmand, S.; Nemati, H. Developed correlations for heat transfer and flow friction characteristics of louvered finned tube heat exchangers. *Int. J. Therm. Sci.* **2018**, *129*, 135–144. [CrossRef]
20. Patankar, S.V.; Liu, C.H.; Sparrow, E.M. Fully Developed Flow and Heat Transfer in Ducts Having Streamwise-Periodic Variations of Cross-Sectional Area. *Trans. ASME. Ser. C, J. Heat Transf.* **1977**, *99*, 180. [CrossRef]
21. Martinez-Espinosa, E.; Vicente, W.; Salinas-Vazquez, M.; Carvajal-Mariscal, I. Numerical Analysis of Turbulent Flow in a Small Helically Segmented Finned Tube Bank. *Heat Transf. Eng.* **2017**, *38*, 47–62. [CrossRef]
22. Awais, M.; Bhuiyan, A.A. Heat and mass transfer for compact heat exchanger (CHXs) design: A state-of-the-art review. *Int. J. Heat Mass Transf.* **2018**, *127*, 359–380. [CrossRef]
23. Qasem, N.A.A.; Zubair, S.M. Compact and microchannel heat exchangers: A comprehensive review of air-side friction factor and heat transfer correlations. *Energy Convers. Manag.* **2018**, *173*, 555–601. [CrossRef]
24. Kristófersson, J.; Vestergaard, N.P.; Skovrup, M.; Reinholdt, L. Ammonia charge reduction potential in recirculating systems—Calculations. In Proceedings of the IIR Ammonia Refrigeration Conference, Ohrid, Macedonia, 11–13 May 2017; pp. 80–87.
25. Kristófersson, J.; Vestergaard, N.P.; Skovrup, M.; Reinholdt, L. Ammonia charge reduction potential in recirculating systems—System benefits. In Proceedings of the IIR Ammonia Refrigeration Conference, Ohrid, Macedonia, 11–13 May 2017; pp. 72–79.
26. Lotfi, B.; Sundén, B. Development of new finned tube heat exchanger: Innovative tube-bank design and thermohydraulic performance. *Heat Transf. Eng.* **2019**, *7632*, 1–27. [CrossRef]
27. Sahiti, N.; Lemouedda, A.; Stojkovic, D.; Durst, F.; Franz, E. Performance comparison of pin fin in-duct flow arrays with various pin cross-sections. *Appl. Therm. Eng.* **2006**, *26*, 1176–1192. [CrossRef]
28. Menter, F.R. Two-equation eddy-viscosity turbulence models for engineering applications. *AIAA J.* **1994**, *32*, 1598–1605. [CrossRef]
29. Kim, M.S.; Lee, J.; Yook, S.J.; Lee, K.S. Correlations and optimization of a heat exchanger with offset-strip fins. *Int. J. Heat Mass Transf.* **2011**, *54*, 2073–2079. [CrossRef]
30. Chimres, N.; Wang, C.C.; Wongwises, S. Optimal design of the semi-dimple vortex generator in the fin and tube heat exchanger. *Int. J. Heat Mass Transf.* **2018**, *120*, 1173–1186. [CrossRef]
31. Chapter 4: Turbulence, 4.16 Near-Wall Treatments for Wall-bounded Turbulent Flows, 4.16.1. In *ANSYS Fluent 19.1 User's Guide*; ANSYS Inc.: Canonsburg, PA, USA, 2019.
32. Shah, R.K.; Sekulic, D.P. *Fundamentals of Heat Exchanger Design*; John Wiley & Sons: Hoboken, NJ, USA, 2003.
33. Webb, R.L.; Kim, N. *Principles of Enhanced Heat Transfer*, 2nd ed.; CRC Press: New York, NY, USA, 2005.
34. Kaminski, U.; Grob, S. Luftseitiger Wärmeübergang und Druckverlust in Lamellenrohr-Wärmeübertragern. *Ki Luft Kaeltetechnik* **2000**, *36*, 13–18.
35. Fraß, K.P.F.; Hofmann, R. *Principles of Finned-Tube Heat Exchanger Design for Enhanced Heat Transfer*, 2nd ed.; WSEAS Press, 2015. Available online: http://www.wseas.org/wseas/cms.action?id=9512 (accessed on 15 September 2019).
36. Shah, R.K.; London, A.L. *Laminar Flow Forced Convection in Ducts*; Academic Press: New York, NY, USA, 1978.

© 2019 by the authors. Licensee MDPI, Basel, Switzerland. This article is an open access article distributed under the terms and conditions of the Creative Commons Attribution (CC BY) license (http://creativecommons.org/licenses/by/4.0/).

Article

Shape Optimization of a Two-Fluid Mixing Device Using Continuous Adjoint

Pavlos Alexias [†,*] and Kyriakos C. Giannakoglou

Parallel CFD & Optimization Unit, School of Mechanical Engineering, National Technical University of Athens, Irroon Polytechniou 9, Zografou 15780, Greece; kgianna@mail.ntua.gr
* Correspondence: p.alexias@engys.com
† Current address: Engys Ltd. Studio 20, Royal Victoria Patriotic Building, John Archer Way, London SW183SX, UK.

Received: 28 November 2019; Accepted: 6 January 2020; Published: 8 January 2020

Abstract: In this paper, the continuous adjoint method is used for the optimization of a static mixing device. The CFD model used is suitable for the flow simulation of the two miscible fluids that enter the device. The formulation of the adjoint equations, which allow the computation of the sensitivity derivatives is briefly demonstrated. A detailed analysis of the geometry parameterization is presented and a set of different parameterization scenarios are investigated. In detail, two different parameterizations are combined into a two-stage optimization algorithm which targets maximum mixture uniformity at the exit of the mixer and minimum total pressure losses. All parameterizations are in conformity with specific manufacturability constraints of the final shape. The non-dominated front of optimal solutions is obtained by using the weighted sum of the two objective functions and executing a set of optimization runs. The effectiveness of the proposed synthetic parameterization schemes is assessed and discussed in detail. Finally, a reduced length mixer is optimized to study the impact of the length of the tube on the device's performance.

Keywords: mixing devices; two-phase flows; shape optimization; continuous adjoint method

1. Introduction

During recent years, there is a growing demand for designing and constructing highly efficient engineering devices and systems. Flow systems are no exception and, thus, the development of optimization tools that improve their performance is of high importance. Computational Fluid Dynamics (CFD) is a highly accurate way to predict the flow behavior within the system and, coupled with an optimization method, consist a both efficient and effective design process.

The optimization of any device starts by defining the objective-function(s) measuring its performance and the design variables. The optimal values of the design variables that minimize (or maximize) the objective function(s) are sought. The minimization (or maximization) of a single objective function, can be carried out using gradient-based methods. These make use of the gradient of the objective function to update the current geometry at the end of each optimization cycle. They converge fast and their cost is exclusively determined by the cost of computing the gradients. There is a variety of methods to compute gradients (finite differences, automatic differentiation [1], complex variables method [2]), with the adjoint [3,4] being the most efficient one, since its cost is independent of the number of design variables. The adjoint method can be developed following the continuous or discrete approach, with both of them having their own advantages and disadvantages. Their main difference relies on whether the differentiation or the discretization of the flow equations comes first. In this paper, the continuous adjoint approach, programmed in the OpenFOAM environment, is used.

When the flow system includes two or more fluids, a multiphase flow model must be used. The way this is formulated greatly depends on the fluid properties, their interaction and their concentrations inside the mixture [5–9]. In this paper, a flow model for two miscible fluids following a Eulerian description is used. This model is suitable for the simulation of flows inside mixing devices which do not contain moving parts. These are motionless structures that blend two or more fluids traveling inside a tube trying to deliver an homogeneous mixture at the exit. They are met in various application fields such as medicine, wastewater treatment and chemistry applications. Their functionality is based on the existence of baffles inside the tubes which force the flow to recirculate enhancing, thus, the mixing process. Apart from delivering uniform flow at the outlet, mixers should have the smallest possible power losses to reduce energy consumption. Several published studies are dealing with the flow simulation in mixing devices [10,11] or with the problem of optimizing them, targeting mixture uniformity at the exit [12–14] and minimum total pressure drop within the device [15,16], though none of them uses the adjoint method, at least to the author's knowledge. In this paper, a method based on the continuous adjoint for a two-phase model is used for the optimization of a static mixing device targeting both the aforementioned objective functions. The continuous adjoint method for this two-phase model has been developed in [17] and is, herein, summarized by presenting the adjoint partial differential equations (PDEs), the adjoint boundary conditions and the gradient expression. For the optimization of the device, the two parameterizations initially presented in [17], namely a node-based and a positional angle one, are used. A significant difference is that, in this paper, the two parameterizations are combined by formulating a two-stage optimization. Over and above, a study of a shorter device is provided to examine the impact of the length on the performance of the device, in view of a forthcoming optimization in which the tube length is an extra design variable.

2. Flow Analysis & Shape Optimization Tools

The flow domain within the static mixing device is enclosed by two inlets (one inlet per incoming fluid), a single outlet (where mixture uniformity is targeted) and the solid walls (including the baffles the shape of which must be optimized). Figure 1 presents the geometry of the mixer, where seven equally distributed baffles are placed inside. In this initial/reference geometry, every second baffle is placed at the same angular position, at 180° shift from its previous/next one.

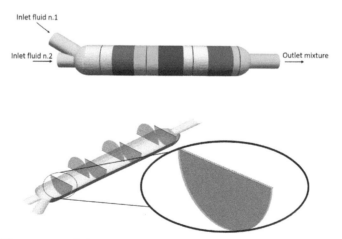

Figure 1. Mixer geometry which comprises of two inlets, one outlet and seven baffles. (**Top**): the mesh blocks across the mixer geometry. Each baffle is associated with a unique mesh region that can be displaced in the peripheral direction ("rotated") independently from the rest ones. (**Bottom**): the set of points (red patch), the coordinates of which comprise the design variables in the NBP.

2.1. Two-Phase Flow Model-Primal Equations

For a laminar flow of two miscible fluids, the flow or primal problem within the optimization loop requires the solution of the flow equations, written in the form [7,9]

$$R^p = -\frac{\partial(\rho v_i)}{\partial x_i} = 0 \tag{1}$$

$$R_i^v = \rho v_j \frac{\partial v_i}{\partial x_j} - \frac{\partial}{\partial x_j}\left[\mu\left(\frac{\partial v_i}{\partial x_j} + \frac{\partial v_j}{\partial x_i}\right)\right] + \frac{\partial p}{\partial x_i} = 0 \quad i=1,2,3 \tag{2}$$

$$R^\alpha = v_i \frac{\partial \alpha}{\partial x_i} - \frac{\partial}{\partial x_j}\left(D\frac{\partial \alpha}{\partial x_j}\right) = 0 \tag{3}$$

where ρ is the mixture density, v_i are the mixture velocity components, p is the static pressure and μ is the mixture dynamic viscosity. In Equation (3), α denotes the volume fraction of the mixture and D the mass diffusivity coefficient. Throughout this paper, repeated indices imply summation. Assuming that both fluids have constant densities (ρ_1 and ρ_2) and constant viscosities (μ_1 and μ_2), the mixture density and viscosity are given by $\rho = \alpha\rho_1 + (1-\alpha)\rho_2$ and $\mu = \alpha\mu_1 + (1-\alpha)\mu_2$.

For the closure of the problem, the following flow or primal boundary conditions are imposed as:

- Inlets (S_I): Fixed incoming velocity components v_i and fixed distributions of the volume fraction α; in specific, Inlet 1 is given $\alpha=1$ (first incoming fluid) and Inlet 2 is given $\alpha=0$ (second fluid). Zero Neumann condition for the static pressure.
- Outlet (S_O): Zero Dirichlet condition for p. Zero Neumann condition for v_i and α.
- Walls (S_W): Zero Dirichlet condition for v_i (no-slip condition). Zero Neumann condition for p and α.

2.2. Shape Parameterization

The shape parameterization defines the variables controlling shape modifications based on the computed (in this work, by the continuous adjoint method) gradients of the objective function. Its selection is important as search based on different shape parameterizations explore different design spaces and, occasionally, lead to different (sub)optimal solutions. The two parameterizations this paper relies on were also used in a previous study, [17], therein independently from each other. Here, the goal is to effectively combine both parameterizations during the optimization to get better performing mixing device configurations. The two parameterizations are:

- Node-Based Parameterization (NBP). The coordinates of each surface node of the selected patches (parameterized walls S_{W_p}) of the computational mesh are the design variables.
- Positional Angle Parameterization (PAP). The angular positions of the baffles across the mixer are used as design variables. This means that, starting from an initial position, the baffles can be placed at different angles inside the mixer without changing either their shapes or their longitudinal positions.

In what follows, the degrees of freedom of the problem are denoted by

$$\vec{b} = (b_1, b_2, ..., b_N) \in \Re^N \tag{4}$$

The above parameterizations will be used in adjoint-based optimization loops for two mixers of different length, without though handling the length as an extra design variable.

2.3. Objective Functions

This paper is dealing with two objective functions, see also [17]. The first one, denoted as F_U, is a measure of the mixture uniformity at the exit. It is defined by

$$F_U = \int_{S_O} v_i n_i \left(\alpha - \frac{\int_{S_O} \alpha dS}{S_O} \right)^2 dS \qquad (5)$$

where n_i is the unit outward normal vector to the outlet boundary. The term into parenthesis in the integral denotes the deviation of the local α from its averaged value over the outlet patch. In a well-mixed flow, F_U tends to zero. The second objective function is related to the (volume flowrate-weighted) total pressure losses occurring between the inlets and the outlet. This is given by

$$F_P = -\frac{1}{2} \int_{S_{I,O}} v_i n_i (p + \frac{1}{2}\rho v_j^2) dS \qquad (6)$$

and should be minimized too.

Since the optimization is carried out using a gradient-based method minimizing a single target function, the two objectives are combined in

$$F = w_1 F_U + w_2 F_P \qquad (7)$$

where w_1 and w_2 are user-defined weights. Practically, these are set as $w_1 = \tilde{w}_1 / F_U^0$ and $w_2 = \tilde{w}_2 / F_P^0$ where F_P^0 and F_U^0 are the values of the objective functions for the reference static mixer geometry. In fact, $\tilde{w}_1$ and $\tilde{w}_2$ are the weights selected by the user. The total derivative of F (expressed, in the general sense, as $F = \int_S F_{S,i} n_i dS$) w.r.t. $\vec{b}$ is

$$\frac{\delta F}{\delta \vec{b}} = \int_{S_I \cup S_O} \frac{\partial F_{S,i}}{\partial \vec{b}} n_i dS + \int_{S_I \cup S_O} \frac{\partial F_{S,i}}{\partial x_k} \frac{\delta x_k}{\delta \vec{b}} n_i dS + \int_{S_I \cup S_O} F_{S,i} \frac{\delta}{\delta \vec{b}} (n_i dS) \qquad (8)$$

In Equation (8), the following identity (see [18])

$$\frac{\delta \Phi}{\delta \vec{b}} = \frac{\partial \Phi}{\partial \vec{b}} + \frac{\partial \Phi}{\partial x} \frac{\delta x}{\delta \vec{b}} \qquad (9)$$

that relates the total (δ) and partial (∂) derivatives of any flow variable Φ, by also involving the mesh sensitivities $\delta x / \delta \vec{b}$, is used.

2.4. Adjoint Equations

To develop the continuous adjoint method that computes the sensitivity derivatives of F w.r.t. $\vec{b}$, the augmented objective function

$$F_{aug} = F + \int_\Omega q R^p d\Omega + \int_\Omega u_i R_i^v d\Omega + \int_\Omega \phi R^a d\Omega \qquad (10)$$

where q, u_i, ϕ are the adjoint pressure, velocities and phase fraction respectively, is defined and differentiated as presented in detail in [17] (for two-phase flows) and [18] (for single-phase flows).

The differentiation of Equation (10) w.r.t. $\vec{b}$ yields

$$\frac{\delta F_{aug}}{\delta \vec{b}} = \frac{\delta F}{\delta \vec{b}} + \int_\Omega \left(q \frac{\partial R^p}{\partial \vec{b}} + u_i \frac{\partial R_i^v}{\partial \vec{b}} + \phi \frac{\partial R^a}{\partial \vec{b}} \right) d\Omega + \int_S (qR^p + u_i R_i^v + \phi R^a) \frac{\delta x_j}{\delta \vec{b}} n_j dS \qquad (11)$$

By using the Green-Gauss theorem to the volume integral of Equation (11), a lengthy development exposed in the aforementioned references provides the adjoint field equations

$$R^q = -\frac{\partial u_i}{\partial x_i} = 0 \quad (12a)$$

$$R^{u_i} = \rho u_j \frac{\partial v_j}{\partial x_i} - \frac{\partial (\rho u_i v_j)}{\partial x_j} - \frac{\partial}{\partial x_j}\left[\mu\left(\frac{\partial u_i}{\partial x_j} + \frac{\partial u_j}{\partial x_i}\right)\right] + \rho \frac{\partial q}{\partial x_i} + \phi \frac{\partial \alpha}{\partial x_i} = 0 \quad i = 1, 2, 3 \quad (12b)$$

$$R^\phi = -\frac{\partial(\phi v_i)}{\partial x_i} - \frac{\partial}{\partial x_j}\left(D\frac{\partial \phi}{\partial x_j}\right) + \rho_\Delta\left(u_i v_j \frac{\partial v_i}{\partial x_j} + v_i \frac{\partial q}{\partial x_i}\right) + \mu_\Delta \frac{\partial u_i}{\partial x_j}\left(\frac{\partial v_i}{\partial x_j} + \frac{\partial v_j}{\partial x_i}\right) = 0 \quad (12c)$$

where $\rho_\Delta = \rho_1 - \rho_2$ and $\mu_\Delta = \mu_1 - \mu_2$. The above set of adjoint field equations is associated with the following set of adjoint boundary conditions:

- Inlets (S_I): Dirichlet condition for the adjoint velocity; in specific the normal component is set to $u_n = -n_i \partial F_{S_{I,i}}/\partial p$ and the tangential ones $u_t^I = u_t^{II} = 0$. Zero-Dirichlet condition for ϕ together with zero-Neumann for q.
- Outlets (S_O): Dirichlet conditions for u_i: $u_n v_n = q$ and $u_t v_n + v \frac{\partial u_t}{\partial n} = 0$. Robin condition for adjoint phase $\phi v_i n_i + D \frac{\partial \phi}{\partial x_j} n_j - \rho_\Delta q v_i n_i = -\frac{\partial (F_i n_i)_{S_O}}{\partial \alpha}$. Zero Neumann condition for q.
- Walls (S_W): Zero Dirichlet condition for u_i. Zero Neumann condition for ϕ and q.

2.5. Sensitivity Derivatives

After satisfying the adjoint field equations and boundary conditions, the resulting terms in (the developed) Equation (11) give the sensitivity derivatives

$$\frac{\delta F}{\delta \vec{b}} = -\int_{S_{W_p}} \left\{\left[-q\rho n_i + \mu\left(\frac{\partial u_i}{\partial x_j} + \frac{\partial u_j}{\partial x_i}\right) n_j\right]\frac{\partial v_i}{\partial x_m} n_m \frac{\delta x_k}{\delta \vec{b}} n_k + \phi D \frac{\partial \alpha}{\partial x_j}\frac{\delta n_j}{\delta \vec{b}} + \phi D \frac{\partial^2 \alpha}{\partial x_k \partial x_j}\frac{\delta x_k}{\delta \vec{b}} n_j\right\} dS \quad (13)$$

Equation (13) is written for a general design variable vector $\vec{b}$, where S_{W_p} is the set of parameterized walls. Working with NBP, applied on the mixer, only the coordinates of points at the top part of each baffle are considered as design variables (Figure 1). By doing so, only the profile of each baffle can be modified whereas its lateral surfaces remain planar. The points are moved only perpendicular to the top part securing this way that each baffle maintains its thickness. Assuming that the tube is aligned with the z-axis, the design vector becomes $\vec{b} = [x_1, x_2, ..., x_M, y_1, y_2, ..., y_M]$ where M is the total number of boundary nodes on the parameterized walls.

With NBP, it is almost mandatory to additionally use a gradient smoothing algorithm and this because any numerical noise in the computed gradient can create irregularities on the surface and lead the optimization loop to diverge. Smoothing, also, allows bigger deformations to be of the surface and, consequently, to converge faster to the optimal solution. A more extensive study on this matter can be found in [19]. For smoothing the gradients, a diffusion-like equation is solved on the surface of the geometry.

$$\tilde{G} - \epsilon \nabla_S^2 \tilde{G} = G \quad (14)$$

where ϵ is a coefficient that defines the intensity of smoothing, $G = \delta F/\delta \vec{b}$ (13) and $\tilde{G}$ is the smoothed sensitivity field which the Equation (14) is solved for. The ∇_S^2 operator is the Laplace-Beltrami operator on the surface of the shape to be modified. Figure 2 demonstrates the different displacements of the top surface of the first baffle when using the non-smoothed and the smoothed gradients. For the adaptation of the internal mesh nodes to the displaced boundaries an inverse distance mesh deformation tool coupled with mesh optimization techniques is used [20].

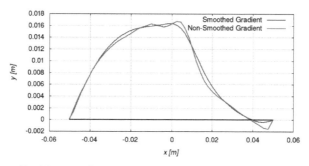

Figure 2. The profile of the top surface of the first baffle at the end of the first optimization cycle (with the NBP) when a non-smoothed (red) or a smoothed (blue) gradient is used. Note that the diameter of the inner cylindrical surface of the tube is 0.1 m.

In case the PAP is used, $\vec{b} = [\theta_1, \theta_2, ..., \theta_B]$, where B is the total number of baffles inside the mixer and θ is the angle of rotation of each baffle. Here, as before, only the top part of the baffle is parameterized. Then, each node on the surface of the baffle can be written in a cylindrical coordinate system as

$$\vec{x}_i = (|\vec{r}_i|\cos\theta, |\vec{r}_i|\sin\theta, z) \tag{15}$$

where $\vec{r}_i$ is a vector pointing from a point on the axis (at the same z) to each node i. Then, the derivative of $\delta F/\delta \theta$ can be computed from Equation (13), by additionally using that

$$\frac{\delta x_k}{\delta b_j} = (-|\vec{r}_i|\sin\theta, |\vec{r}_i|\cos\theta, 0) \tag{16}$$

While changing the positional angle of each baffle, the latter needs to slide along the inner wall of the mixer, which requires either a complicated mesh adaptation algorithm or to redesign the geometry on the CAD system. To avoid this, each baffle is associated with a different mesh block, as shown in Figure 1. By doing so, all cylindrical blocks can be displaced in the peripheral direction independently from each other. This alleviates the need to slide the baffles along the wall and adapt the mesh accordingly.

During the solution, consecutive mesh blocks are communicating by interpolating each discrete field v_i, p, a over their non-matching interfaces (in the PAP). The same holds also for the adjoint fields u_i, q and ϕ. The interpolation is done between two interfaces A and B that are geometrically identical, but with different distribution of nodal positions (Figure 3). To do this, for each face f_i over the interface A, all the faces f_j belonging to B which it overlaps with are tracked down. For each f_j, the relative weight contribution is calculated as $W_{i,j} = S_{f_i}/S_{f_j}$, where S is the surface area of each face. This way, the interpolated value of a variable Φ from interface B to A becomes as $\Phi_A = \sum_j^K W_{i,j}\Phi_j$ with K being the total number of overlapping faces.

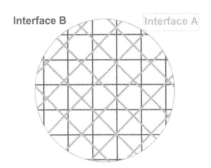 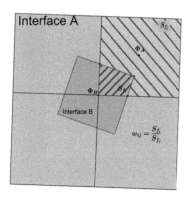

Figure 3. Field interpolation patterns between two non-matching interfaces, for use in the PAP-based optimization.

Both parameterizations can be used as stand-alone tools (as was the case in [17]), but can also be combined into a single workflow. This way, the top surface of the baffle can be deformed and, at the same time, the positional angles of the baffle can be changed. In this paper, the two parameterization schemes are combined in three different optimization scenarios:

1. The first scenario with two consecutive stages in which the NBP is used until convergence is reached and, afterwards, the PAP takes over starting from the converged solution of the first stage.
2. The opposite two-stage scenario, in which the PAP (until convergence) is used and, afterwards, the NBP takes over.
3. A scenario in which both parameterizations are used simultaneously (coupled usage) at each optimization cycle.

2.6. Optimization Workflow

The optimization workflow is as follows:

1. The primal (1) and, then, the adjoint (12) equations are solved.
2. Based on the primal and adjoint fields, the sensitivity derivatives are computed using Equation (13).
3. In the NBP (only), gradients are smoothed out through Equation (14).
4. The design variables are updated using steepest descent as $\vec{b}^{new} = \vec{b}^{old} - \eta G_{old}$, where G_{old} denotes the previously computed (possibly smoothed) gradient.
5. The mesh is then adapted to the change of the design variables. In the NBP, an inverse distance morphing method is use to adapt the rest of the mesh nodes, the coordinates of which are not design variables. In the PAP, each mesh region is peripherally displaced following the baffle "rotation".
6. The process is repeated starting from Step 1 until the convergence criterion is satisfied.

3. Results

The static mixer consists of a main 0.77 m long cylindrical body (tube) with inner diameter of 0.1 m, two inlets, one outlet and comprises seven baffles as shown in Figure 1. The baffles have semi-circular shapes, every second of which is placed exactly at the same angle; two consecutive baffles are placed with 180° difference (reference geometry). Their role is to force the flow to recirculate for increasing mixing. The longitudinal positions of the baffles are listed in Table 1, with number 1 corresponding to the baffle closest to the two inlets.

Table 1. Longitudinal positions of the baffles across the static mixer.

Baffle No.	1	2	3	4	5	6	7
Longitudinal Position [m]	0.05	0.125	0.2	0.275	0.350	0.425	0.5

Two different fluids enter the device, from a different inlet each, with known mass flow rates (0.29 and 0.26 kg/s, respectively). The first (second) fluid properties are: density 1500 kg/m^3 (1300 kg/m^3) and kinematic viscosity 1.5×10^{-5} m^2/s (1.3×10^{-5} m^2/s).

The Reynolds number of the flow based on the mean values of viscosity and mass flow rate of the two fluids is ~450 and, thus, the simulation is performed assuming laminar flow. An unstructured hexahedral-based mesh with approximately 200 K cells is generated. This mesh is sufficiently refined, as further increase in the mesh size has no impact on the values of the objective functions. Two optimization cases with the same flow properties, though with different degrees of freedom, have been studied in [17]. Recall that the purpose of this paper is to combine the parameterizations proposed in [17] and, by doing this, get even better solutions for the same objectives.

In this section, all plots presenting the computed optimal solutions use the objective functions F_U (Equation (5)) and F_P (Equation (6)) divided by the (fixed) volume flow rate; no special symbols for the so-modified functions are used.

3.1. Optimization Scenario 1

In Scenario 1, a two-stage optimization process is performed. In the first stage, the optimization is based on the NBP, running until convergence; this is then followed by a second optimization stage based on the PAP. In this second stage, the shapes (and, of course, the longitudinal positions) of the baffles computed in the first stage are retained but the baffles are allowed to change their angular positions. Figure 4 demonstrates the fronts of non-dominated solutions that result upon completion of each optimization stage. Six different value-sets of weights ($\tilde{w}_1, \tilde{w}_2$) are used as in the caption of Figure 4. An important observation, is that the front of non-dominated solutions at the end of the second stage clearly dominates over all the members of the first stage front. The way the flow develops inside the mixer is presented in Figure 5 which illustrates the velocity streamlines coloured by the phase fraction.

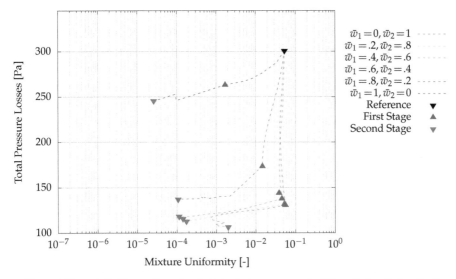

Figure 4. Scenario 1. Fronts of non-dominated solutions computed at the end of each stage for the two-stage optimization approach using six different sets of weight values.

Figure 5. Scenario 1. Velocity streamlines coloured by the phase fraction for the reference geometry (**top-left**), the optimized geometry with $\tilde{w}_1 = 1, \tilde{w}_2 = 0$ (**top-right**), and that with $\tilde{w}_1 = 0, \tilde{w}_2 = 1$ (**bottom**).

The geometries of the non-dominated solutions are shown in Figure 6. Also, Figure 7 demonstrates the phase fraction over the outlet plane for each value-set of weights for all the non-dominated solutions. It is noticeable that, for high $\tilde{w}_2$ values, the NBP tries to remove material from the baffles in order to avoid increasing the total pressure losses caused as a consequence of intensive flow recirculation. This, of course, has a negative impact on the mixing of the two fluids. In addition, in the extreme case where $\tilde{w}_1 = 1$ and $\tilde{w}_2 = 0$, the PAP turns all the baffles towards the same side of the mixer and makes "space" for the fluid to flow with the least resistance to its motion. On the other hand, when higher weighting values are associated with F_U, the profile of the baffles acquires a "wavy" shape which improves the mixing performance. In addition, by optimizing the angular positions of the baffles, these are placed so as to redirect the vorticity vector of the recirculation causing increased flow mixing. The way the flow develops in the devices corresponding to the two extreme points of the front (the ones with either $\bar{w}_1 = 0$ or $\bar{w}_2 = 0$) is presented in Figure 5.

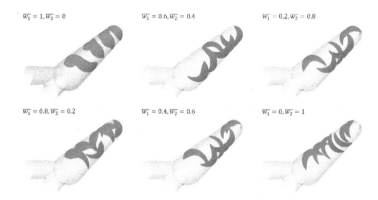

Figure 6. Scenario 1. Optimal baffle shapes for each set of weights.

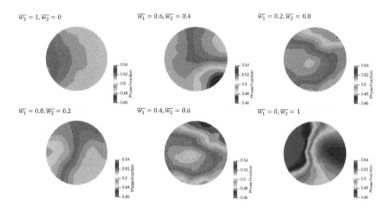

Figure 7. Scenario 1. Final distribution of the phase fraction at the outlet for each set of weights.

Figure 8 demonstrates the shape change of the first and the last baffle during the two-stage optimization process for all the value-sets of weights.

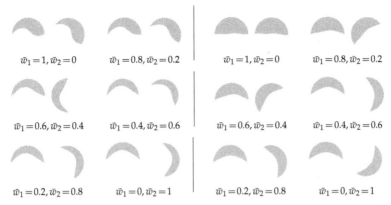

Figure 8. Scenario 1. Optimized shape and angular position of the first (**left**, in each pair of plots) and the last (**right**) baffle, for each value-set of weights.

3.2. Optimization Scenario 2

In this scenario, again a two-stage optimization is carried out, this time in reverse order though. This means that the PAP (starting from the same reference geometry as in the previous section) runs first until convergence, followed by the NBP optimization stage. In the second stage, the angular positions of the baffles are fixed (to their values computed in the first stage). Figure 9 demonstrates the fronts of non-dominated solutions of the two optimization stages. An interesting difference resulting from the comparison of the front of non-dominated solutions in Figure 9 with the one obtained from Scenario 1, is that the first stage gives greater improvements in the objective functions (creating a more extended front) compared to the first stage of Scenario 1. In addition, the second stage contributes less to the overall reduction in the objective function values.

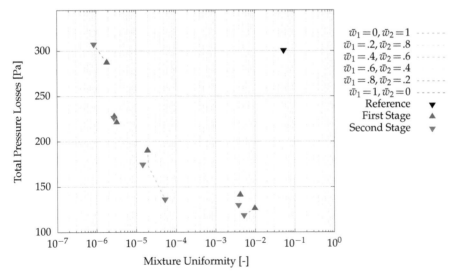

Figure 9. Scenario 2. Fronts of non-dominated solutions computed at the end of each stage using six different sets of weight values.

Figure 10 presents the final baffle geometries using the two-stage optimization for the six value-sets of weights. Here, similarly to Scenario 1, the same behaviour is observed depending on the weights of the objective functions. If emphasis is laid on F_U, alternating baffles with "wavy" profiles must be used; in contrast, if F_P is given priority the baffles become shorter and are placed towards the same side of the mixer walls.

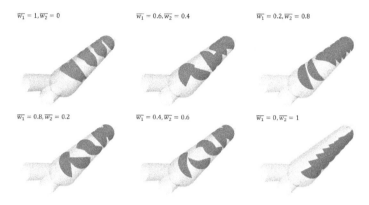

Figure 10. Scenario 2. Perspective views of the optimal baffle shapes and peripheral locations for each set of weights.

3.3. Optimization Scenario 3

In the third optimization scenario, the same two parameterization techniques are used but, this time, not as the synthesis of two successive stages, as in Scenarios 1 and 2. In this case, a "coupled" optimization is used according to which, in each optimization cycle, both parameterizations are simultaneously used. Figure 11 presents the front of non-dominated solutions computed using this coupled optimization workflow together with the fronts resulted by the two two-stage optimizations (Scenarios 1 and 2). As it can be seen from Figure 11, all the optimization approaches are contributing to the final front with four members each. The solutions obtained using Scenario 1 (first NBP, then PAP) dominate in the area of small F_P values. In contrast, the solutions for Scenario 2 (first PAP, then NBP) perform better in the area of small F_U values. Finally, Scenario 3 ("coupled") has a wider spread across the front contributing the two extreme points to the "Front of Fronts" (namely the points with the smallest F_U and F_P value).

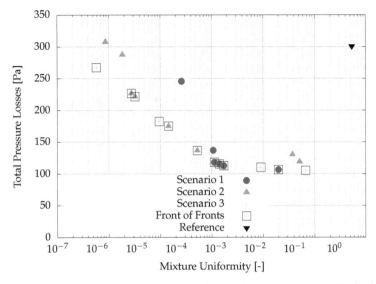

Figure 11. Fronts of non-dominated solutions for all the optimization scenarios. The final front of non-dominated solutions (empty squares) from all optimizations ("Front of Fronts") as well as the reference configuration are included.

3.4. Optimization of a Reduced Length Mixer

To further investigate how different geometric characteristics impact the performance of the mixing device, the length of the mixer is reduced together with the number of the baffles. The goal is to measure and compare (with the previous scenarios) the performance of the reduced length tube when using the "coupled" approach (Scenario 3). The purpose of choosing the "coupled" approach is because it has been shown that is offers the most wide-spread non-dominated front compared to other approaches. In detail, the length of the new tube is $0.54m$ and the number there are only four baffles. The diameter of the mixer and the characteristics of the two fluids remain the same. The longitudinal positions of the baffles are given in Table 2. Figure 12 presents the mixer geometry coloured by the mesh regions that each baffle belongs to.

Figure 12. Geometry of the mixer with reduced length and number of baffles.

Table 2. Reduced Length Mixer. Longitudinal positions of the four baffles.

Baffle No.	1	2	3	4
Longitudinal Position [m]	0.05	0.125	0.2	0.275

By solving the primal equations, the computed values of F_U and F_P for the reduced length mixer (reference configuration) are presented in Table 3 together with the ones computed for the regular length mixer (reference configuration, too). As expected, due to the smaller length and the reduced number of baffles, a higher drop in F_P is observed at the expense, of course, of worst F_U values.

Table 3. Reduced Length Mixer. Objective function values for the reference mixer geometries of two different lengths.

	F_P	F_U
Regular Length Mixer	300.69 Pa	0.0538
Reduced Length Mixer	221.07 Pa	0.0734

Running six optimization problems using the "coupled" approach (as in Scenario 3) with the same value-sets of weights, the non-dominated front of optimal solution is computed and depicted in Figure 13 together with the objective values of the reference (reduced length) geometry. In the same graph, the non-dominated front of the regular tube geometry is included too. It can be seen that the optimal solutions of the reduced length mixer are dominating in the low F_P region extending the range of the front of non-dominated solutions towards this area. Finally, Figure 14 demonstrates the phase fraction distribution at the outlet patch of the mixer for the three different optimization scenarios and for the reduced length mixer (computed with Scenario 3). The demonstrated results concern optimizations done targeting only the F_U. As it can be seen in Figure 14, Scenario 3 delivers an almost perfectly homogeneous mixture, whereas the reduced length mixer has noticeable differences from all the regular length scenarios.

For all scenarios, a single optimization run convergences in around 6 CPU hours using 4 Intel Core i7-6800K 3.40 GHz processors. The optimization turnaround time can be significantly reduced by switching to a much faster quasi-Newton method based on approximations to the objective function; this, however, affects only the computational cost and not the quality of the obtained results.

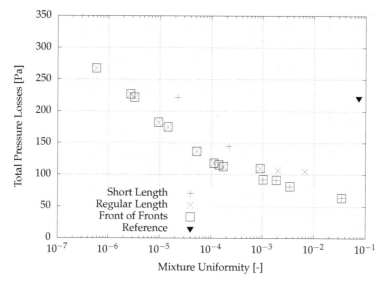

Figure 13. Reduced Length Mixer. Fronts of non-dominated solutions for the reduced length mixer, using Scenario 3. The final front of non-dominated solutions ("Front of Fronts") from all optimizations is demonstrated (empty squares).

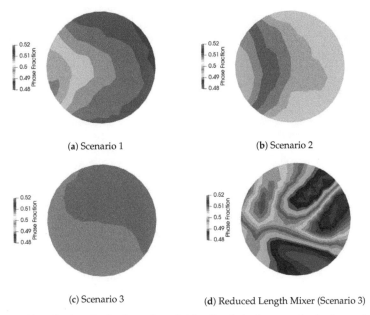

Figure 14. Phase fraction distribution at the outlet for all optimization scenarios for the regular mixer and Scenario 3 for the reduced length mixer. The weights used are $\vec{w}_1 = 1$ and $\vec{w}_2 = 0$. Note that scale is narrowed down to [0.48, 0.52] to better illustrate the differences among them.

4. Conclusions

The optimization of two static mixers with different lengths and number of baffles was carried out using the continuous adjoint method. Different combinations of parameterizations were tried out, with each one contributing differently into the computed front of non-dominated solutions.

The performed studies show that the consecutive combination of two parameterizations during the optimization is beneficial as it allows either to further improve the optimal solution(s) obtained with only one parameterization (see also [17]) or to converge to other non-dominated solutions, enriching this way the final front. More specifically, Scenario 1 (first NBP, then PAP) produced better results in terms of F_P, whereas Scenario 2 (first PAP, then NBP) performed better in the area of low F_U values. Also, when the two parameterizations were simultaneously used, a new set of well-spread non-dominated solutions, without favoring a particular objective, came out. In an additional study, the length of the tube and the number of baffles were reduced, offering this way a significant drop in total pressure losses, compromising on the mixture uniformity, compared to the regular length mixer.

Author Contributions: Conceptualization, P.A. and K.C.G.; Data curation, P.A.; Methodology, P.A. and K.C.G.; Supervision, K.C.G.; Writing—original draft, P.A.; Writing—review & editing, P.A. and K.C.G. All authors have read and agreed to the published version of the manuscript.

Funding: This research received funding by the European Union HORIZON 2020 Framework Programme for Research and Innovation under Grant Agreement No. 642959.

Acknowledgments: Parts of this work have been conducted within the IODA project (http://ioda.sems.qmul.ac.uk), funded by the European Union HORIZON 2020 Framework Programme for Research and Innovation under Grant Agreement No. 642959.

Conflicts of Interest: The authors declare no conflict of interest.

References

1. Rall, L. *Automatic Differentiation: Techniques and Applications*; Springer: New York, NY, USA, 1981.
2. Martins, J.R.R.A.; Sturdza, P.; Alonso, J.J. The Complex-step Derivative Approximation. *ACM Trans. Math. Softw.* **2003**, *29*, 245–262. [CrossRef]
3. Pironneau, O. *Optimal Shape Design for Elliptic Systems*; Springer: Berlin/Heidelberg, Germany, 1984.
4. Jameson, A. Aerodynamic design via control theory. *J. Sci. Comput.* **1988**, *3*, 233–260. [CrossRef]
5. Hirt, C.; Nichols, B. Volume of fluid (VOF) method for the dynamics of free boundaries. *J. Comput. Phys.* **1981**, *39*, 201–225. [CrossRef]
6. Brennen, C. *Fundamentals of Multiphase Flow*; Cambridge University Press: Cambridge, UK, 2005.
7. Ishii, M.; Hibiki, T. *Thermo-Fluid Dynamics of Two-Phase Flow*; Springer: New York, NY, USA, 2011.
8. Drew, D.A. Mathematical Modeling of Two-Phase Flow. *Annu. Rev. Fluid Mech.* **1983**, *15*, 261–291. [CrossRef]
9. Manninen, M. *On the Mixture Model for Multiphase Flow*; Technical Research; Centre of Finland: Espoo, Finland, 1996.
10. Zhang, M.; Zhang, W.; Wu, Z.; Shen, Y.; Chen, Y.; Lan, C.; Cai, W. Comparison of Micro-Mixing in Time Pulsed Newtonian Fluid and Viscoelastic Fluid. *Micromachines* **2019**, *10*, 262. [CrossRef] [PubMed]
11. Zhang, M.; Cui, Y.; Cai, W.; Wu, Z.; Li, Y.; chen Li, F.; Zhang, W. High Mixing Efficiency by Modulating Inlet Frequency of Viscoelastic Fluid in Simplified Pore Structure. *Processes* **2018**, *6*, 210. [CrossRef]
12. Hanada, T.; Kuroda, K.; Takahashi, K. CFD geometrical optimization to improve mixing performance of axial mixer. *Chem. Eng. Sci.* **2016**, *144*, 144–152. [CrossRef]
13. Regner, M.; Östergren, K.; rdh, C.T. Effects of geometry and flow rate on secondary flow and the mixing process in static mixers—A numerical study. *Chem. Eng. Sci.* **2006**, *61*, 6133–6141. [CrossRef]
14. Rudyak, V.; Minakov, A. Modeling and Optimization of Y-Type Micromixers. *Micromachines* **2014**, *5*, 886–912. [CrossRef]
15. Hirschberg, S.; Koubek, R.; F.Moser; Schock, J. An improvement of the Sulzer SMX^{TM} static mixer significantly reducing the pressure drop. *Chem. Eng. Res. Des.* **2009**, *87*, 524–532. [CrossRef]
16. Song, H.; Han, S.P. A general correlation for pressure drop in a Kenics static mixer. *Chem. Eng. Sci.* **2005**, *60*, 5696–5704. [CrossRef]
17. Alexias, P.; Giannakoglou, K.C. Optimization of a static mixing device using the continuous adjoint to a two-phase mixing model. In *Optimization and Engineering*; Springer: Berlin, Germany, 2019.
18. Papoutsis-Kiachagias, E.M.; Giannakoglou, K.C. Continuous Adjoint Methods for Turbulent Flows, Applied to Shape and Topology Optimization: Industrial Applications. *Arch. Comput. Methods Eng.* **2016**, *23*, 255–299. [CrossRef]
19. Jameson, A.; Vassberg, J. Studies of Alternate Numerical Optimization Methods Applied to the Brachistochrone Problem. In Proceedings of the OptiCON '99 Conference, Newport Beach, CA, USA, 14–15 October 1999.
20. Alexias, P.; de Villiers, E. Gradient Projection, Constraints and Surface Regularization Methods in Adjoint Shape Optimization. In *Evolutionary and Deterministic Methods for Design Optimization and Control With Applications to Industrial and Societal Problems*; Andrés-Pérez, E., González, L.M., Periaux, J., Gauger, N., Quagliarella, D., Giannakoglou, K., Eds.; Springer International Publishing: Cham, Switzerland, 2019; pp. 3–17.

© 2020 by the authors. Licensee MDPI, Basel, Switzerland. This article is an open access article distributed under the terms and conditions of the Creative Commons Attribution (CC BY) license (http://creativecommons.org/licenses/by/4.0/).

Article

Valve Geometry and Flow Optimization through an Automated DOE Approach

Micaela Olivetti [1,*], Federico Giulio Monterosso [1,*], Gianluca Marinaro [2], Emma Frosina [2] and Pietro Mazzei [2]

1 R&D and Engineering Department, Omiq Srl, 20135 Milan, Italy
2 Industrial Engineering Department, University of Naples Federico II, 80125 Naples, Italy; gianluca.marinaro@unina.it (G.M.); emma.frosina@unina.it (E.F.); pie.mazzei@studenti.unina.it (P.M.)
* Correspondence: olivetti@omiq.it (M.O.); monterosso@omiq.it (F.G.M.)

Received: 4 December 2019; Accepted: 28 January 2020; Published: 30 January 2020

Abstract: The objective of this paper is to show how a completely virtual optimization approach is useful to design new geometries in order to improve the performance of industrial components, like valves. The standard approach for optimization of an industrial component, as a valve, is mainly performed with trials and errors and is based on the experience and knowledge of the engineer involved in the study. Unfortunately, this approach is time consuming and often not affordable for the industrial time-to-market. The introduction of computational fluid dynamic (CFD) tools significantly helped reducing time to market; on the other hand, the process to identify the best configuration still depends on the personal sensitivity of the engineer. Here a more general, faster and reliable approach is described, which uses a CFD code directly linked to an optimization tool. CAESES® associated with SimericsMP+® allows us to easily study many different geometrical variants and work out a design of experiments (DOE) sequence that gives evidence of the most impactful aspects of a design. Moreover, the result can be further optimized to obtain the best possible solution in terms of the constraints defined.

Keywords: optimization; valves; computational fluid dynamic (CFD); CAESES®; SimericsMP+®

1. Introduction

It is well known that main and pilot stage valves, adopted in hydraulic circuits, have different performance requirements. Typically, main stage valves have high efficiency with adequate bandwidth and power, while pilot ones have rapid transient response and are stable and robust when facing external disturbances. When the power required by the pilot stage comes directly from the main line, pressure affects dynamic behavior and stability, making it difficult to tune the system to respond correctly to all pressure loads [1,2]. For this reason, different solutions are generally used to separate the two stages and make them as independent as possible.

The present study shows a technique to optimize a pilot operated distributor solenoid/hydraulic controlled valve. The presented modeling technique is based on the adoption of two tools, the optimization tool CAESES® (Friendship Systems AG, Postdam, Germany) and a commercial computational fluid dynamics (CFD) code: SimericsMP+® (Simerics Inc.®, Bellevue, WA, USA).

This approach is faster than the one already presented by the authors [1,2] and can be applied to several geometries for the study of the components' internal fluid patterns.

Several examples of valve optimization are available in literature: some of them focused on the fluid dynamic, others on structural aspects [1–11].

Optimization tools and techniques are quite common in structural analysis, as they are used to reduce local stresses or to improve topology of mechanical parts.

For example, Park et al. [12] proposed an approach based on a traditional structural optimization, which identifies the best combination of geometrical parameters to improve the product's performance and to save material. This paper presents a framework that performs the integration between commercial CAD–CAE software. This approach reduces the time for solving computation-intensive design optimization problems so that designers are free from monotonous repetitive tasks. The results show that the proposed method facilitates the structural optimization process and reduces the computing cost compared to other approaches.

Regarding the fluid dynamic aspect, the main problem is to identify how the fluid behaves inside the component. Some examples of fluid dynamic optimization can be found in literature [5–7].

Manring et al. [7,8] modeled a spool–valve to study the flow forces acting on the valve spool. In other scientific papers, the same authors showed the experimental investigation carried out on hydraulic spool valves to measure the pressure transient force action on the valve's spool. The importance of optimizing fluid dynamic forces in modeling and testing approaches was demonstrated by these studies.

Zardin et al. [9] studied valves for mobile applications via a lumped parameter approach. They proposed an innovative design procedure to optimize valve design. The technique involves dedicated simulations to analyze the main critical issues regarding a cartridge valve. Models and simulations were used to define a methodology for designing a new valve. The optimized valve satisfies the requirements and adapts well to the necessities of operating at higher flow and pressure levels without compromising performances.

A useful tool to understand the flow behavior inside a component is three-dimensional computational fluid dynamics, a collection of different numerical techniques that allow to solve the Navier–Stokes equations.

Unfortunately, a main obstacle to implement optimization studies in fluid-dynamics analysis is, still today, computational cost. Furthermore, the setup of such projects typically requires three different tools to interact efficiently: a parametric geometry modeler (CAD), a computational fluid dynamics (CFD) solver and an optimization tool.

Tonomura et al. [13] showed a methodology for the optimization of a microdevice. Even if this component is not in the fluid power field, the approach used could be easily adopted in many research sectors. Authors studied a specific part inside the component using computational fluid dynamics (CFD). Then, a CFD-based optimization method was proposed for the design of plate-fin microdevices. With this approach, the optimal shape was designed almost automatically.

Corvaglia et al. [10] showed an interesting study on a load sensing proportional valve. The valve was modelled using two 3D CFD numerical approaches. The models were validated in terms of flow rate and pressure drop for different positions of the main spool by means of specific tests. This paper brought to evidence the reliability of the CFD models in evaluating the steady-state characteristics of valves with complex geometry.

Salvador et al. [11] adopted a computational fluid dynamics (CFD) approach to design hydraulic components such as valves by inexpensively providing insight into flow patterns, potential noise sources and cavitation. They demonstrated the relevance of the geometric characteristics on the performance. A modification of the geometry in the piston exit leads, for example, to different vortex structures and helps reduce vibrations and forces on the piston.

As mentioned before, Frosina et al. [1,2,8] already studied the valves' fluid-dynamics in order to analyze flow forces, pressures distribution and velocity behavior. All these studies were performed using 1D and 3D CFD modeling approaches depending on the application. Studies have demonstrated the accuracy of the developed methodologies and showed good agreement with experimental data. Geometric parameters were characterized and consequently modified systematically. The three-dimensional model's results, like velocity behavior and pressure distribution, allowed the authors of the study to optimize the valve geometry without losing any of the valve's performance. In this context, it would have been very advantageous to have access to an automated procedure that could drastically reduce the project duration.

For the project described in this article, just two tools were used: CAESES® (an optimization tool with integrated parametric geometry modelling capabilities) and SimericsMP+®, a commercial CFD solver. This approach greatly reduced the set-up effort and allowed for a leaner and more efficient project layout.

The objective of this work is to show how the shape of a valve ports can be automatically modified, without the use of an external CAD tool, and simulated to obtain the best performing geometry in just a few hours.

The design taken into consideration for the optimization is the geometry of a four-way hydro-piloted valve for industrial applications. In particular, the shape of two ports of the valve was optimized in order to obtain the highest possible mass flux at an imposed pressure drop.

The study began from a baseline geometry, tested with the CFD tool, from which the optimization started.

In the following paragraphs, the integration between the optimizer and the CFD tool as well as the results obtained will be described.

2. Materials and Methods

The DSP10 valve by Duplomatic MS S.p.A. (Parabiago-MI, Italy) was the object of the optimization study (Figure 1).

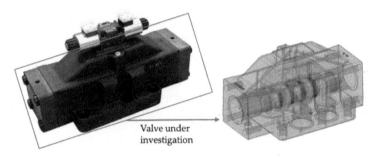

Figure 1. Valve under investigation.

It is worth noting that a good overall agreement between CFD studies conducted with SimericsMP+® on similar Duplomatic MS S.p.A. valves and experimental tests performed at the Industrial Engineering Department of the University of Naples, Federico II are reported in different publications (e.g., [1,8]).

For optimization purposes, the valve was simulated with fixed spool position so that only ports P and A (in blue in Figure 2) were connected through the spool port recesses (green in Figure 2).

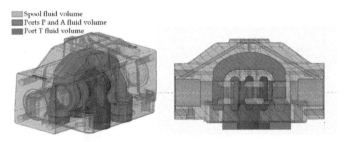

Figure 2. Ports A and P (blue) and spool caves (green).

The volume wetted by the oil (Figure 3) was extracted with a CAD tool, and an STL file was exported to be used within the CFD code SimericsMP+® (developed by Simerics Inc.®, Bellevue, WA, USA)).

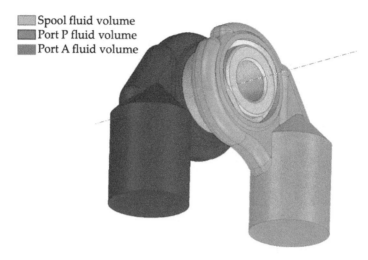

Figure 3. Fluid volumes of the valve.

In the performed study, the SimericsMP+® tool was chosen as a general purpose CFD software that numerically solves the fundamental conservation equations of mass, momentum and energy as described below [14,15].

For the purposes of the study, some simplifications were considered, such as a stationary domain, a steady state flow and an isothermal flow. Given these approximations, some terms of the equations written below are disregarded by the solver during the run.

Mass conservation:

$$\frac{\partial}{\partial t}\int_{\Omega(t)} \rho d\Omega + \int_{\sigma} \rho(v - v_\sigma)\cdot n d\sigma = 0 \quad (1)$$

Momentum conservation:

$$\frac{\partial}{\partial t}\int_{\Omega(t)} \rho v d\Omega + \int_{\sigma} \rho((v-v_\sigma)\cdot n)v d\sigma = \int_{\sigma} \widetilde{\tau}\cdot n d\sigma - \int_{\sigma} pn d\sigma + \int_{\Omega} f d\Omega \quad (2)$$

Energy conservation:

$$\frac{\partial}{\partial t}\left[\rho\left(u + \frac{v^2}{2} + gz\right)\right] + \nabla\left[\rho v\left(h + \frac{v^2}{2} + gz\right)\right] + \nabla Q - \nabla(T_d v) = 0 \quad (3)$$

in which

- $\Omega(t)$ is the control volume,
- σ is the control volume surface,
- n is the surface normal pointed outwards,
- ρ is the fluid density,
- p is the pressure,
- f is the body force,
- v is the fluid velocity,
- v_σ is the surface motion velocity.

$\tilde{\tau}$, the shear stress tensor, is a function of the fluid viscosity μ and of the velocity gradient. For a Newtonian fluid, this is given by the following Equation (4),

$$\tau_{ij} = \mu\left(\frac{\partial u_i}{\partial x_j} + \frac{\partial u_j}{\partial x_i}\right) - \frac{2}{3}\mu\frac{\partial u_k}{\partial x_k}\delta_{ij} \tag{4}$$

where u_i ($i = 1,2,3$) is the velocity component and δ_{ij} is the Kronecker delta function.

The software implements mature turbulence models, such as the standard $k - \varepsilon$ model and Re-Normalization Group (RNG) $k - \varepsilon$ model [16]. These models have been available for more than a decade and are widely demonstrated to provide good engineering results. The standard $k - \varepsilon$ model, used for the simulations presented in this paper is based on the following two equations:

$$\frac{\partial}{\partial t}\int_{\Omega(t)} \rho k d\Omega + \int_\sigma \rho((v - v_\sigma)n)k d\sigma = \int_\sigma \left(\mu + \frac{\mu_t}{\sigma_k}\right)(\nabla k n)d\sigma + \int_\Omega (G_t - \rho i e)d\Omega \tag{5}$$

$$\frac{\partial}{\partial t}\int_{\Omega(t)} \rho i e d\Omega + \int_\sigma \rho((v - v_\sigma)n)\varepsilon d\sigma = \int_\sigma \left(\mu + \frac{\mu_t}{\sigma_\varepsilon}\right)(\nabla i e n)d\sigma + \int_\Omega \left(c_1 G_t \frac{\varepsilon}{k} - c_2 \rho \frac{\varepsilon^2}{k}\right)d\Omega \tag{6}$$

with $c_1 = 1.44$, $c_2 = 1.92$, $\sigma_k = 1$, $\sigma_\varepsilon = 1.3$; where σ_k e σ_ε are the turbulent kinetic energy and the turbulent kinetic energy dissipation rate Prandtl numbers.

The turbulent kinetic energy, k, is defined as:

$$k = \frac{1}{2}(v' \cdot v') \tag{7}$$

with v' being the turbulent fluctuation velocity, and the dissipation rate, ε, of the turbulent kinetic energy is defined as:

$$\varepsilon = 2\frac{\mu}{\rho}\left(S'_{ij}S'_{ij}\right) \tag{8}$$

in which the strain tensor is:

$$S'_{ij} = \frac{1}{2}\left(\frac{\partial u'_i}{\partial x_j} + \frac{\partial u'_j}{\partial x_i}\right) \tag{9}$$

with u_i' ($i = 1,2,3$) being components of v'.

The turbulent viscosity μ_t is calculated by:

$$\mu_t = \rho C_\mu \frac{k^2}{ie} \tag{10}$$

with $C_\mu = 0.09$.

The turbulent generation term $G_{t\,can}$ be expressed as a function of velocity and the shear stress tensor as:

$$G_t = -\overline{\rho u'_i u'_j}\frac{\partial u'_i}{\partial x_j} \tag{11}$$

where $\tau'_{ij} = \overline{\rho u'_i u'_j}$ is the turbulent Reynolds stress, which can be modelled by the Boussinesq hypothesis:

$$\tau'_{ij} = \mu_t\left(\frac{\partial u_i}{\partial x_j} + \frac{\partial u_j}{\partial x_i}\right) - \frac{2}{3}\left(\rho k + \frac{\partial u_k}{\partial x_k}\right)\delta_{ij} \tag{12}$$

The valve fluid volume was meshed with the SimericsMP+® grid generator (Figure 4).

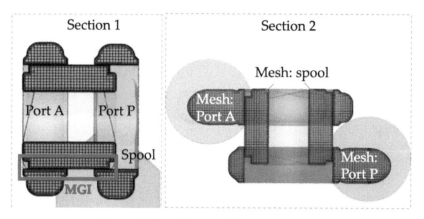

Figure 4. Grid seen from two different section planes.

SimericsMP+® uses a body-fitted binary tree approach [14,15]
This type of grid is accurate and efficient because:

- The parent–child tree architecture allows for an expandable data structure with reduced memory storage;
- Binary refinement is optimal for transitioning between different length scales and resolutions within the model;
- Most cells are cubes, which is the optimum cell type in terms of orthogonality, aspect ratio and skewness, thereby reducing the influence of numerical errors and improving speed and accuracy;
- It can be automated, greatly reducing the set-up time.

In the configuration considered for the optimization, the spool is fixed in the position that allows the flux from Port P to Port A. The fluid volumes of the ports and the spool were meshed separately and were then connected via an implicit interface.

The SimericsMP+® mismatched grid interface (MGI, see Figure 5) is a very efficient implicit algorithm that identifies the overlap areas and matches them without interpolation. During the simulation process, the matching area is treated no differently than an internal face between two neighboring cells in the same grid domain.

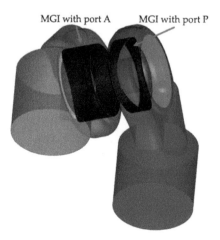

Figure 5. Mismatched grid interface (MGI) between the spool and both ports.

Thanks to this approach, the solution becomes very robust, quick and accurate.

The DSP10 valve, the object of the study, was optimized at the most typical condition with a pressure difference of 5 bar.

The CFD model of the considered valve portion consists of 911,150 cells (Figure 4).

The following boundary conditions were applied (Figure 6):

- Fluid: oil at 45 °C (constant)
- Oil kinematic viscosity: 4.42×10^{-5} [m^2/s] = 44.2 cSt
- Oil density: 876 [kg/m^3]
- Inlet, Port P: fixed static absolute pressure 50 bar
- Outlet, Port A: fixed static absolute pressure of 45 bar

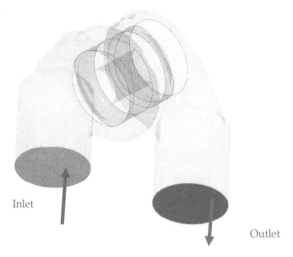

Figure 6. Boundary conditions on Port A and P.

A static analysis with turbulence was performed on the model. Run time for this analysis was 14 min on an 8 cores Intel Core i7, 3.10 GHz processor with 32 Mb RAM.

In this configuration, a baseline CFD analysis was performed, to be used as reference during the optimization process.

As previously indicated, the optimization process was driven by CAESES®.

CAESES® stands for "CAE System Empowering Simulation" and its ultimate goal is to design optimal flow-exposed products [17]. Starting from a baseline geometry, it is possible within CAESES® to modify the geometry, using different strategies and imposing constraints and parameters to obtain a set of geometries and boundary conditions that will be treated as a design of experiments (DOE) set.

The strategies used for the geometry modifications are:

- Fully parametric modeling: It allows the user to build the geometry from scratch in CAESES®, using a proprietary "Meta Surface technology". This technology gives the possibility of modifying the built-in geometry in all possible ways (Figure 8).
- Partially parametric modeling: It lets the user import existing geometries and morph or deform these geometries. This means that the original geometry can be "distorted and modified" using a sort of surrounding grid, with control points that drive the geometry modifications (Figure 7).

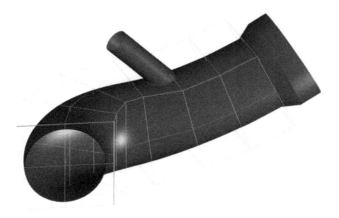

Figure 7. Partially parametric modelling.

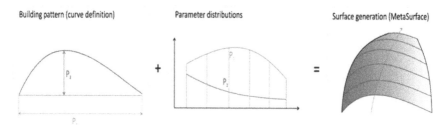

Figure 8. Fully parametric modelling.

Once the geometric strategy was chosen, CAESES® calculated all the possible shapes within the defined constraints and calculated a DOE sequence for the valid geometrical solutions.

The DOE sequence can also take into account variations of the boundary conditions, but the performed study was only based on geometrical modifications.

The ports of the valve, the object of the study, were modelled in CAESES® using the "fully parametric modelling" approach. The "partial parametric modeling" approach was used for other parts of the model (spool and other ports), although these parts have not been included in this phase of the project.

This means that the original geometry was rebuilt in CAESES® and different geometrical modifications of the valve ports A and B were taken into consideration.

CAESES® allows the user to select the geometry control parameters that are deemed relevant for the problem.

In the specific case, nine parameters for each port were identified:

- Box height (Figure 9)
- Box rotation
- Box shift
- Cap height
- Cap rotation
- Cylinder height
- Cylinder inclination
- Outer radius (Figure 10)
- Outer fillet

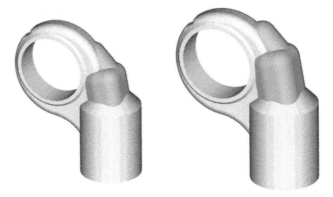

Figure 9. Box height variation. Left original, right max modification.

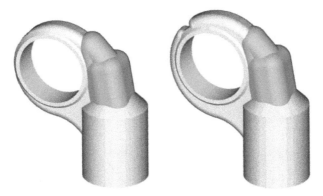

Figure 10. Outer radius variation. Left original, right max modification.

For example, in Figure 9, the box height modification is shown. In Figure 10 the outer radius variation is illustrated.

Not all the control parameters were used for the optimization: a DOE sequence generated with a Sobol algorithm identified four modifications for each port for a totally of eight design variables and 90 variants. In Table 1 these values are resumed.

Table 1. The eight design variables with their upper and lower values.

Parameter	Lower Value	Upper Value	Initial Value
Box shift for Port A	−2.5 [mm]	−1.8 [mm]	−2 [mm]
Box rotation for Port A	5 [°]	10 [°]	10 [°]
Outer circle radius for Port A	1.45 [mm]	1.6 [mm]	1.5 [mm]
Outer fillet radius for Port A	10 [mm]	35 [mm]	30 [mm]
Box shift for Port P	−2.5 [mm]	−1 [mm]	−1.1 [mm]
Box rotation for Port P	5 [°]	10 [°]	9 [°]
Outer circle radius for Port P	1.45 [mm]	1.6 [mm]	1.482 [mm]
Outer fillet radius for Port P	10 [mm]	35 [mm]	34.61 [mm]

Two variables were monitored in CAESES®: Port A and Port P volumes were monitored not to exceed predefined values.

The objective of the optimization was to maximize the mass flow rate of the valve at a fixed pressure drop.

As the DOE sequence was defined, the CFD simulations for the 90 variants were performed with SimericsMP+®.

The great advantage of using CAESES® is that the code drives all the process automatically; this means that CAESES® generates the geometry that has to be tested on the base of the "design variables".

CAESES® creates the STL file that is used by SimericsMP+® to generate the mesh. SimericsMP+® is then run in batch and generates the new mesh, sets up the simulation and solves the case.

The results from SimericsMP+® are read, via a .txt file, from CAESES®, that evaluates the obtained mass flux value.

Figure 11 illustrates the process scheme:

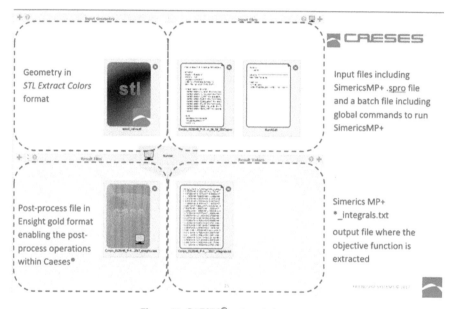

Figure 11. CAESES® automated process.

The CFD analyses were performed on all the 90 design variants.

Considering a mean simulation time of 15 min for SimericsMP+®' shared memory parallel solver on a single processor, eight cores workstation, the whole DOE sequence calculation took 22.5 h; less than one day.

3. Results

3.1. Baseline Computational Fluid Dynamics (CFD) Results

Results obtained on the valve are shown in Figures 12–15. In particular, Figure 12 shows the pressure distribution on the walls of the fluid domain of the baseline geometry. It is clear that pressure was distributed according to the boundary conditions applied.

Other significative results are shown with two cross sectional views of the fluid domain: Figure 13 shows the pressure distribution, while Figure 14 is representative of the velocity behavior inside the domain.

Flux behavior inside the ports is also described with streamlines colored with the velocity magnitude in Figure 15.

The mass flux obtained with the baseline geometry (13.47 [kg/s]) was used as starting value for the optimization.

The objective was therefore to find the maximum possible mass flux compatible with the prescribed constraints.

Figure 12. Pressure distribution on Port A and P walls, baseline geometry.

Figure 13. Pressure distribution on sections of Port A and Port P, baseline geometry.

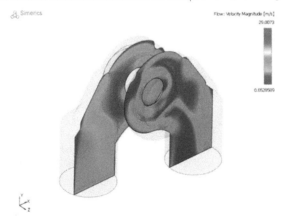

Figure 14. Velocity distribution on sections of Port A and Port P, baseline geometry.

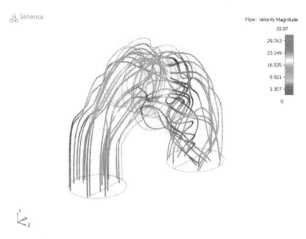

Figure 15. Streamlines in the valve, baseline geometry.

3.2. Optimizaion Results

At the end of the DOE sequence solution process, CAESES® provides a detailed table of all the data used in the calculations. For each simulated design, the corresponding geometric characteristics as well as calculations results are provided. In this specific project, as previously mentioned, 90 design variants were tested. A chart mapping 67 solutions versus the obtained flow rate can be visualized in Figure 16.

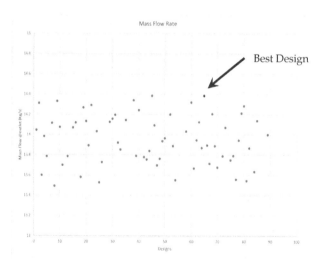

Figure 16. Design of experiments (DOE) results, in terms of mass flow at the outlet of the valve.

The remaining 23 solutions are not shown in the chart because of calculations failure. This means either that the simulation didn't end correctly or that the geometry could not be built with the prescribed parameters. All the 90 ports geometries had volumes within the limits requested, so that they could be contained in the original valve compartment. Ports volumes were monitored during the optimization process, even if they were not considered as a strict constraint.

The best mass flux obtained was 14.38 [kg/s] that, compared with the baseline result of 13.47 [kg/s], provided a 6.8% flow-rate increment.

Table 2 sums up the geometric parameter values of the best solution, in comparison to the baseline geometry.

Table 2. Best design geometric values compared to the baseline geometry.

Parameter	Baseline	Optimized
Box height for Port A	−2 [mm]	−1.871 [mm]
Box rotation for Port A	10 [°]	7.617 [°]
Outer circle radius for Port A	1.5 [mm]	1.592 [mm]
Outer fillet radius for Port A	30 [mm]	17.62 [mm]
Box height for Port P	−1.1 [mm]	−1.949 [mm]
Box rotation for Port P	9 [°]	8.555 [°]
Outer circle radius for Port P	1.482 [mm]	1.5847 [mm]
Outer fillet radius for Port P	34.61 [mm]	14.88 [mm]
Volume Port A	158,242 [mm^3]	175,369 [mm^3]
Volume Port B	158,967 [mm^3]	178,111 [mm^3]

At the end of DOE sequence calculation, a parameter sensitivity analysis was performed to determine which parameter had the greatest influence on the mass flux.

Figure 17 shows the influence of the outer circle radius for Port A and P on the mass flux of the valve:

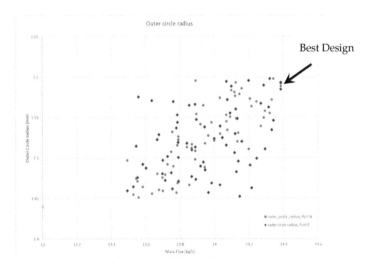

Figure 17. Outer circle radius impact on the optimization.

This parameter was the most effective in changing the mass flow value. In fact, small changes in the outer radius diameter provided a significant change in the mass flux: a 0.13 mm increment corresponded approximately a 1 kg/s mass flux increase.

The process continued with a 2-level "Tsearch" optimization, starting from the best Sobol sequence design. "Tsearch" optimization is an optimization method based on the local tangent minimum and is aimed at improving the solution within the neighborhood of the selected design.

The T-Search method was originally proposed by Hilleary in 1966 [18]. It combines smaller steps and larger moves through the design space (a pattern search) and directly handles inequality constraints (see [19] for an elaboration). Mathematically speaking, it is a gradient-free method, but it comes up with probing moves not dissimilar to gradient directions.

Results of the 2-level "Tsearch" optimization were very close to the best geometry obtained with the Sobol design of experiment sequence: they provided a further 2% increment in the mass flow rate of the valve.

Figure 18 shows the results obtained with the T-Search optimization.

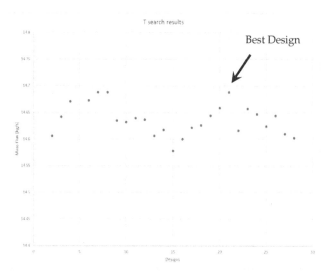

Figure 18. 2-level T-Search result.

In Table 3, best design results against baseline are compared.

Table 3. T-Search Optimization Results.

Parameter	Baseline	Best Design
Mass Flow rate [kg/s]	13.47	14.70
Outer radius Port A [mm]	1.5	1.595
Outer radius Port P [mm]	1.5	1.595
Volume Port A [m^3]	0.000168	0.000176
Volume Port P [m^3]	0.000170	0.000180

Optimized Geometry

The final geometry obtained is illustrated in Figure 19. In Figure 20, a comparison between the baseline geometry and the optimized geometry is shown.

Figure 19. Final optimized geometry.

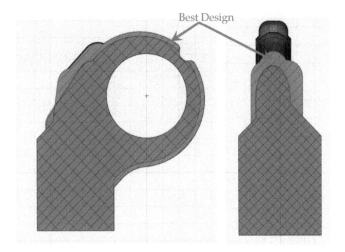

Figure 20. Geometry comparison.

As Figures 19 and 20 show, the main differences are on outer radius and box height, both in Port P and Port A.

3.3. CFD Results on Optimized Geometry

Results of CFD analysis are shown in Figures 21–24. The set-up of the analysis is the same described previously. In the images, the variables were set with the same scale as in the baseline simulation, for an easier comparison.

Figure 21. Pressure distribution on Port A and P walls, optimized geometry.

Figure 22. Pressure distribution on sections of Port A and Port P, optimized geometry.

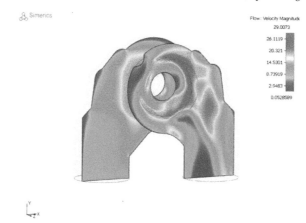

Figure 23. Velocity distribution on sections of Port A and Port P, optimized geometry.

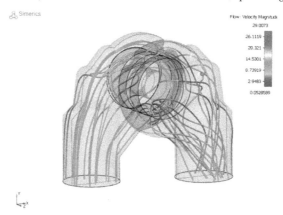

Figure 24. Streamlines in the valve, optimized geometry.

4. Discussion

This project shows how valve design can be virtualized and automated, provided that efficient and reliable software tools are available.

The advantage of this approach is the possibility of studying many different geometry variations, simply defining the parameters that are to be investigated. New geometries are automatically generated by CAESES® and then evaluated by Simerics MP+®. Answers can be obtained in very short time, also with different optimization techniques.

Optimization in this case was based on a two-step strategy. The first one, based on the Sobol design of experiment sequence, provides a geometry that let the valve increase the mass flux by about 7%; the second, a T-Search method optimization, further adjusted the geometry to increase the mass flux by another 2%. The overall process allowed for a 9% improvement in the mass flux. The ports' outer radius turned out to be the parameter that mostly influences the result.

Modifying this parameter allows an increase in the ports' volumes, and consequently a higher mass flux can be obtained.

However, larger ports' dimensions might be risky in terms of decelerating the fluid flowing in the valves. CFD results on the new geometry show that this is not the case, as the fluid velocities are not reduced significantly and are comparable with the baseline geometry velocities. The optimized geometry has also an advantage in terms of fluid behavior. Figure 25 shows a comparison of the velocity vectors distribution on the outlet ports for the baseline and optimized geometries: the vortex at the outlet port of the second geometry is significantly reduced.

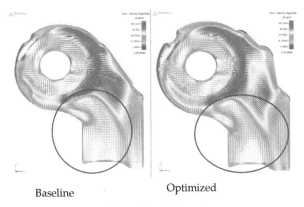

Baseline   Optimized

Figure 25. Comparison of the outlet vortex.

In Figure 26, another advantage of the larger outer radius is shown. Velocity distribution in the outer circumference is smoother in the new geometry and enters with an angle better aligned to the port exit section.

Experimental tests, carried out by Duplomatic MS S.p.A. at the Industrial Engineering Department at the University of Naples Federico II, show that the shape obtained by the optimization process are reliable, as expected from the conducted study.

It is worth noting that the authors conducted different studies on similar spool valves in order to achieve better performances. The Industrial Engineering Department proceeded to optimize the ports geometry with a traditional trial and error approach using laboratory testing and CFD. The results obtained in seven months are aligned to the results obtained with the optimization project performed with CAESES® and SimericsMP+®.

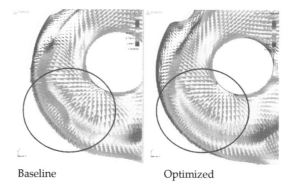

Baseline   Optimized

Figure 26. Comparison of vortex in the outer circumference.

Although these two different approaches reached the same conclusions, two main points should be noted: First of all, the project timeline; seven months for the trial and error approach, one day for the automated approach.

Secondly, the methodology; the trial and error approach can be highly affected by engineer specific expertise while the automated approach is neutral in this respect and somehow free to investigate even apparently unreasonable solutions.

For both methods, the CFD simulation is an essential tool that helps understanding the behavior of the fluid inside the ports, either to find a new solution or to understand the reason for a solution being the optimal one.

5. Conclusions

A fast and reliable methodology to optimize the shape of the ports of a spool valve in order to obtain a higher mass flux was described. SimericsMP+® and CAESES® were used for this project.

Through these tools, an optimized geometry was automatically identified in a very short time. The advantage of the approach is that no parametric CAD tool is needed as CAESES® directly handles the automated process, including geometric modifications, simulations set up and run.

Moreover, a fast and reliable CFD simulation software, as Simerics MP+®, is necessary, as it accelerates the process to obtain the best geometry.

The conducted study also gave evidence of the fact that an optimizer is useful to identify the parameters that mostly influence the objective. Meanwhile, coupled with an efficient CFD solver, it allows investigation of the physics of the problem and determination of the sensitivity of the parameters.

Author Contributions: Conceptualization, M.O.; data curation, G.M. and P.M.; formal analysis, G.M. and P.M.; methodology, M.O. and E.F.; project administration, M.O. and F.G.M.; resources, G.M.; supervision, F.G.M.; Writing—original draft, M.O.; Writing—review & editing, M.O. All authors have read and agreed to the published version of the manuscript.

Funding: This research was developed as part of a PhD program supported by the Italian Government and the MIUR (Ministry of Education, Universities and Research).

Acknowledgments: We would like to thank Ceyhan Erdem e Mike Saroch at Friendship Systems AG for providing outstanding technical support on this project.

Conflicts of Interest: The authors declare no conflict of interest.

Abbreviations/Nomenclature

CAD   Computer Aided Design
CFD   Computational Fluid Dynamics
DOE   Design of Experiments
MGI   Mismatched Grid Interface
STL   Stereolithography
RNG   Re-Normalization Group

References

1. Frosina, E.; Buono, D.; Senatore, A.; Stelson, K.A. A Mathematical Model to Analyze the Torque Caused by Fluid-Solid Interaction on a Hydraulic Valve. *J. Fluids Eng.* **2016**, *138*, 061103. [CrossRef]
2. Frosina, E.; Buono, D.; Senatore, A.; Stelson, K.A. A modeling approach to study the fluid dynamic forces acting on the spool of a flow control valve. *J. Fluids Eng* **2016**, *139*, 011103–011115. [CrossRef]
3. Ivantysynova, M.; Ivatysyn, J. *Hydrostatic Pumps and Motors'*; Vogel Buchverlag: Würzburg, Germany, 2001.
4. Fitch, E.C.; Homg, I.T. *Hydraulic Component Design and Selection'*; Bardyne Inc.: Stillwater, OK, USA, 2004.
5. Akers, A.; Gassman, M.; Smith, R. *Hydraulic Power System Analysis'*; CRC Press: Boca Raton, FL, USA, 2006.
6. Senatore, A.; Buono, D.; Frosina, E.; Pavanetto, M.; Costin, I.I.; Olivetti, M. Improving the control performance of a Proportional spool valve, using a 3D CFD modeling. In Proceedings of the IMECE04 2014 ASME International Mechanical Engineering Congress and Exposition, Montreal, Canada, 14–20 November 2014.
7. Manring, N.D. Modeling Spool-Valve Flow Force. In Proceedings of the ASME 2004 International Mechanical Engineering Congress and Exposition, Anaheim, CA, USA, 13–19 November 2004; pp. 23–29.
8. Manring, N.D.; Zhang, S. Pressure Transient Flow Forces for Hydraulic Spool Valves. *J. Dyn. Sys. Meas. Control.* **2011**, *134*, 034501. [CrossRef]
9. Zardin, B.; Borghi, M.; Cillo, G.; Rinaldini, C.A.; Mattarelli, E. Design Of Two-Stage On/Off Cartridge Valves For Mobile Applications. *Energy Procedia* **2017**, *126*, 1123–1130. [CrossRef]
10. Corvaglia, A.; Altare, G.; Finesso, R.; Rundo, M. Computational fluid dynamics modelling of a load sensing proportional valve. Proceeding of the ASME-JSME-KSME 2019 8th Joint Fluids Engineering Conference, San Francisco, CA, USA, 28 July–1 August 2019.
11. Plau-Salvador, G.; Gonzalez-Altozano, P.; Arviza-Valverde, J. Three-dimensional Modeling and Geometrical Influence on the Hydraulic Performance of a Control Valve. *ASME J. Fluids Eng.* **2008**, *130*, 0111021–0111029. [CrossRef]
12. Park, H.-S.; Dang, X.-P. Structural optimization based on CAD–CAE integration and metamodeling techniques. *Comput. -Aided Des.* **2010**, *42*, 889–902. [CrossRef]
13. Tonomura, O.; Tanaka, S.; Noda, M.; Kano, M.; Hasebe, S.; Hashimoto, I. CFD-based optimal design of manifold in plate-fin microdevices. *Chem. Eng. J.* **2004**, *101*, 397–402. [CrossRef]
14. SimericsMP+ User Manual. Available online: https://www.simerics.com/simerics-valve/ (accessed on 29 January 2020).
15. Ding, H.; Visser, F.C.; Jiang, Y.; Furmanczyk, M. Demonstration and Validation of a 3D CFD Simulation Tool Predicting Pump Performance and Cavitation for Industrial Applications. *J. Fluids Eng.* **2011**, *133*, 011101. [CrossRef]
16. Launder, B.E.; Spalding, D.B. The numerical computation of turbulent flows. *Comput. Methods Appl. Mech. and Eng.* **1974**, *3*, 269–289. [CrossRef]
17. CAESES® User Manual. Available online: https://www.caeses.com/products/caeses/ (accessed on 29 January 2020).
18. Hilleary, R.R. *The Tangent Search Method of Constrained Minimization*; Naval Postgraduate School: Monterey, CA, USA, 1966.
19. Birk, L.; Harries, S. *OPTIMISTIC-Optimization in Marine Design*; Mensch & Buch Verlag: Berlin, Germany, 2003.

© 2020 by the authors. Licensee MDPI, Basel, Switzerland. This article is an open access article distributed under the terms and conditions of the Creative Commons Attribution (CC BY) license (http://creativecommons.org/licenses/by/4.0/).

Article

Experimental Investigation of Finite Aspect Ratio Cylindrical Bodies for Accelerated Wind Applications

Michael Parker [1] and Douglas Bohl [2,*]

[1] Nel Hydrogen, Wallingford, CT 06492, USA; parkermj1@gmail.com
[2] Department of Mechanical and Aeronautical Engineering, Clarkson University, Potsdam, NY 13699, USA
* Correspondence: dbohl@clarkson.edu

Received: 26 January 2020; Accepted: 13 February 2020; Published: 17 February 2020

Abstract: The placement of a cylindrical body in a flow alters the velocity and pressure fields resulting in a local increase in the flow speed near the body. This interaction is of interest as wind turbine rotor blades could be placed in the area of increased wind speed to enhance energy harvesting. In this work the aerodynamic performance of two short aspect ratio (AR = 0.93) cylindrical bodies was evaluated for potential use in "accelerated wind" applications. The first cylinder was smooth with a constant diameter. The diameter of the second cylinder varied periodically along the span forming channels, or corrugations, where wind turbine blades could be placed. Experiments were performed for Reynolds numbers ranging from 1×10^5 to 9×10^5. Pressure distributions showed that the smooth cylinder had lower minimum pressure coefficients and delayed separation compared to the corrugated cylinder. Velocity profiles showed that the corrugated cylinder had lower peak speeds, a less uniform profile, and lower kinetic energy flux when compared to the smooth cylinder. It was concluded that the smooth cylinder had significantly better potential performance in accelerated wind applications than the corrugated cylinder.

Keywords: finite aspect ratio cylinder; accelerated wind; wind energy

1. Introduction

The power that can be extracted from the wind is primarily driven by three factors, the cross-sectional area that is being used to capture the wind, the velocity of the captured wind, and the power coefficient of the turbine blades. This relationship is described by:

$$P = \frac{1}{2}\rho\pi R^2 U_{fs}^3 C_{pow} \tag{1}$$

Here C_{pow} is the coefficient of power for the wind turbine, U_{fs} is the free stream wind speed, and R is the rotor radius [1]. Note that in this work C_p is used to define the pressure coefficient rather than the coefficient of power. One can, therefore, increase the harvested power by increasing the blade radius, the wind speed, or the coefficient of power, which is representative of the aerodynamic efficiency of the wind turbine. Equation (1) indicates that the rotor area is a strong driver of the power of a wind turbine. This has led to increased rotor sizes, especially for commercial scale wind turbines. However, wind turbine sizes are ultimately bounded by structural limitations and other practical considerations such as the need to transport parts. Power is also a function of the coefficient of power of the system which has an upper limit defined by the Betz limit (59.3%) for free wind turbines, limiting potential gains through increased aerodynamic efficiency.

Equation (1) indicates that a wind turbine's extracted power has a cubic relationship with wind speed. This leads to the strategy of increasing the velocity of the wind at the rotor plane to increase the power extraction. Increased wind speed is typically accomplished by siting turbines in locations with high wind speeds or by increasing the tower height. This, however, limits the number of economically viable siting locations. Alternately, one could attempt to modify the local wind stream to achieve higher velocities at the rotor plane. This approach is attractive in that relatively small changes in the local wind speed can lead to significant increases in harvested energy.

"Accelerated wind" is a general term for such strategies and is normally accomplished by adding a structure near the rotor to locally increase the flow velocity. The most common example is seen in diffuser augmented wind turbines (DAWT). A DAWT's structure lowers the pressure downstream of the blades to draw a greater mass of air through the rotor plane and thus generate more power than a similarly sized horizontal axis wind turbine (HAWT) [1–10]. A less explored accelerated wind concept is to place rotors near structures that increase the local wind speed. Examples of this include building augmented wind turbines [11–13] and specially designed tower structures [14–16]. This study explores the concept of placing wind turbine rotors next to a cylindrical structure, see Figure 1. The cylindrical structure serves to act as both the wind turbine tower and a method to increase the velocity at the rotor plane.

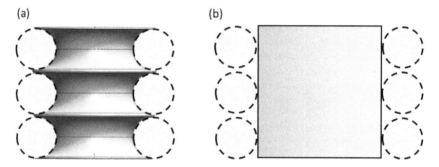

Figure 1. Conceptual views of (**a**) original Optiwind 150 kW wind turbine design. (**b**) Conceptual smooth cylinder design. Rotors shown schematically by dotted circles. In application, direction of wind would be into the page so that turbine blades are at ±90° with respect to the wind direction.

Duffy and Jaran [12] reported on what they named a "toroidal accelerator rotor platform" (TARP). The TARP concept used a toroidal channel around the outside of a cylinder to accelerate the wind into rotor blades that were mounted in the channel. The TARP was intended to be either an add-on attachment to grain silos, water towers, etc. or as a standalone structure. This concept was extended to the WARP, or wind amplified rotor platform, consisting of a number of stacked TARP modules [13]. A prototype was built and briefly tested in Belgium; however, a viable commercial product does not exist today.

A similar concept, the Optiwind "Accelerator Platform", shown in Figure 1a, formed the motivation for the current study. This concept was a finite span (aspect ratio, AR = 0.93) corrugated circular cylinder where the rotor blades would also sit in isolated channels. The channels were conceived of as aerodynamic structures to direct the wind into the wind turbine blades, which would also isolate the wind turbine blades from each other. This is strategically different from DWATs in that the channels were not intended to be traditional diffusers, but more specifically as flow directors. Flow acceleration was provided by the surface curvature. This concept was the motivation for the first model used in the current study. The second model, shown in Figure 1b, was a smooth circular cylinder with the same aspect ratio and a diameter equal to the outer diameter of the corrugated cylinder. The location of the rotor placement for both designs is indicated by the dashed circles. Both models were intended to accelerate the wind prior to entering rotors; however, the smooth cylinder lacked the "flow directing" channels as shown. The high level goal of the project was to design a mid-range, scalable wind turbine for the renewable energy market [15]. In both cases the number of stack turbines, three shown in Figure 1 and used for testing, could be chosen arbitrarily depending on the power requirements.

This work details experiments performed on the two tower models: a 1:80 scale model of the Optiwind Accelerator Platform (i.e., the "corrugated cylinder model") and a smooth circular cylinder with the same aspect ratio as the corrugated model. The surface pressures and tangential velocity, V_θ (r), profiles were acquired experimentally for both platform models. The location and magnitude of the minimum pressure coefficient ($C_{p,min}$) and the mean flow velocities were used as metrics in determining the effectiveness of the potential designs for accelerating the flow. The minimum pressure coefficient, $C_{p,min}$, serves as one basis for discussion of the performance in this work in two ways. First, the location of the $C_{p,min}$ is indicative of where the surface flow curvature has changed and the flow is no longer following the surface shape. It can be used to determine if the separation point has moved forward or aft between cases. Second, the magnitude of $C_{p,min}$ correlates with the increase in flow velocity, and it can again be used comparatively between cases.

The motivation for this work was to assess the potential of the two shapes for accelerated wind applications. The goal of the current work was to compare, in a quantitative manner, the flow around low aspect ratio cylinders with smooth and corrugated surfaces. Within the larger project, the results of this study were used to down select the platform shape and guide the continued design/development of the prototype wind accelerator platform within the larger project. It is acknowledged that the presence of rotors, which were not investigated in this work, would change the flow conditions around both the corrugated and smooth cylinders. This effect is the subject of future studies for the following reasons. First, rotors are typically designed to provide a specific pressure drop that optimizes the power extraction. Because this study was used to down select the platform geometry, the rotors have not yet been designed. Second, this study provides a canonical case comparing short aspect ratio cylinders with and without surface corrugations. These conditions, without the rotors, therefore represent the upper limit on the potential performance.

2. Materials and Methods

2.1. Experimental Models and Facility

The corrugated cylinder was constructed using stereolithography as a 1:80 scaled model of the proposed Optiwind Accelerator Platform design with three rows of corrugations, as shown in Figure 2. The key dimensions of the model were as follows: major diameter, D_{maj} = 0.269 m; minor diameter D_{min} = 0.164 m; length, L = 0.249 m; and the aspect ratio based on the major diameter L/D_{maj} = 0.93. Pressure taps were added circumferentially along the minor diameter (i.e., in the "valley" of the channel) of the model and along the corrugation walls, as shown in Figure 3, for all three channels. The surface of the corrugated cylinder was sanded to smooth the steps in the surface resulting from the stereolithography fabrication.

A smooth cylinder with the same diameter as the major diameter and aspect ratio of the corrugated cylinder provided a second potential platform design as well as a canonical baseline reference case. The smooth cylinder was fabricated from a 0.27 m diameter PVC pipe cut to the same length as the corrugated cylinder, which provided an aspect ratio equal to the platform model, L/D = 0.93. Pressure taps were machined into the cylinder at the mid-height.

Experiments were performed in the Clarkson University High Speed Aerodynamic Wind Tunnel. The tunnel is an open loop tunnel with a 1.2 × 0.9 × 1.8 m long test section. The tunnel blockage due to the models was 6.7% based on the major diameter and length of the corrugated cylinder model. Experimental flow speeds ranged from U_{fs} = 10 to 50 m/s. The turbulence level of the tunnel free stream was measured via hotwire anemometry to be approximately 1.2% within the velocity range investigated. The Reynolds number was computed based on the major diameter of the corrugated cylinder and the test section free stream speed. Experimental Reynolds numbers covered a range of Re = $U_{fs} D_{maj}/\nu = 1.9 \times 10^5$ to 8.9×10^5 for this study. It is noted that the Reynolds number for the full scale device was expected to be nominally 50–100×10^5. The upper end of the experimental Reynolds number was limited in this work by the flow facility (i.e., cross sectional area of the test section and maximum flow speed). While this was approximately an order of magnitude lower than the device Reynolds number that motivated the study, the results show a decreasing dependence on the Reynolds number, and the results were expected to be qualitatively similar and therefore informative. The lowest Reynolds number was investigated to allow for Reynolds number trends to be investigated. Both models were placed in the wind tunnel with a 0.15 m vertical offset from the bottom floor of the test section, as shown in Figure 4. The tops and bottoms of the models were closed.

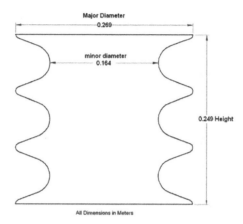

Figure 2. Schematic of corrugated cylinder model. All dimensions are in meters.

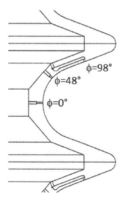

Figure 3. Schematic of side wall pressure taps.

 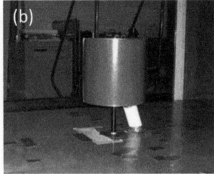

Figure 4. Experimental set-up for the (a) corrugated and (b) smooth models.

2.2. Pressure Measurements

The models contained pressure taps with 1 mm diameter openings at the surface starting at the leading edge ($\theta = 0°$) and extending around the diameter of the models in 10° increments. Stainless steel tubing (1.58mm outer diameter, 1.32 mm inner diameter) was pressed into each tap to allow for connection with the pressure transducer via Tygon tubing. The corrugated cylinder model also had rows of pressure taps along the channel walls at $\phi = 42°$ and 98° up from the horizontal as shown in Figure 3. Pressure surveys were conducted using an Omega model PX653-10BD5V pressure transducer with a ±2.5 kPa range. Data were acquired with a National Instruments PCI-6024E 12 bit A/D card. Each pressure measurement consisted of 96,000 data points at a sampling rate of 2400 Hz. A ScaniValve solenoid controller was used to index through the model pressure taps sequentially after sampling at a given location was completed. Uncertainty in the pressure measurements was estimated to be 0.025 kPa, which corresponded to an uncertainty level in the reported pressure coefficients of $Cp = \pm0.03$.

The pressure data reported in this work are the average surface pressure values in non-dimensional form. The pressure coefficient, C_P, was calculated using:

$$C_p = \frac{\overline{P}_S - \overline{P}_{S,Tunnel}}{\frac{1}{2}\rho U_{fs}^2} \quad (2)$$

where $\overline{P}_S$ is the average pressure at a tap location and $\overline{P}_{S,Tunnel}$ is the static pressure in the test section upstream of the model. The dynamic pressure of the free stream was measured using a pitot-static probe upstream of the models.

2.3. Hot-Wire Measurements

Velocity surveys were taken around the models using a DISA type 55M01 Constant Temperature Anemometer (CTA) with a DANTEC 55P14 single-wire probe. The hot-wire sensor utilized a 5 μm tungsten wire with a 1mm active length. Data were acquired with a National Instruments PCI-6024E 12 bit A/D card. Hot-wire data were sampled at 12,000 Hz for 30 s to provide 360,000 measurements. The hot-wire probes were pre- and post-calibrated to ensure the sensor did not drift during use. The uncertainty in the hot-wire measurements was estimated to be $\pm0.058\ V_\theta/U_{fs}$. Velocity profiles were performed by traversing the probe radially outward at 13 different angular locations around the model over the range of $\theta = 0°$ to 180° in 15° increments.

3. Results

3.1. Review of Flow Around Circular Cylinders

The inviscid solution to flow around a 2D circular cylinder [17] provides the limiting case for the current study. The solid line in Figure 5 shows the surface pressure distribution, while the dashed line shows the associated tangential surface velocity, V_θ, distribution for the inviscid flow around a cylinder. The leading edge of the cylinder is defined as $\theta = 0°$. In the absence of viscosity, the peak speed around a circular cylinder is $V_\theta/U_{fs} = 2$ and is located at $\theta = 90°$. This corresponds to a minimum pressure coefficient of $C_{p,min} = -3$ at the same angular location. Equation (1) indicates that a factor of 2 increase in the wind speed would result in a factor of 8 potential increase in the harvestable power at the surface of the cylinder. It is worth noting that the actual increase in power would be lower than a factor of 8 as V_θ decreases in the radial direction even for the inviscid case; therefore, the actual increase in power would depend on the radius of the wind turbine blades.

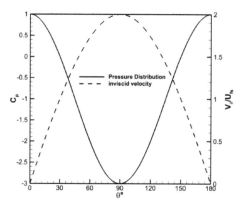

Figure 5. Surface velocity and pressure distribution for inviscid flow around a circular cylinder.

Past results for circular cylinders without end effects [18–21] show that for Reynolds numbers of practical importance, the flow separates and forms a wake downstream of the cylinder. The location of the separation generally begins at about $\theta \approx 70°$ and moves aft (i.e., increasing θ) with the increasing Reynolds number to an angle of $\theta \approx 120°$, as shown in Figure 6. These results also show that the wake region becomes smaller with the increasing Reynolds number. The Re = 8.5×10^5 case shown in Figure 6 does not follow the Reynolds number trends as $C_{p,min}$ is significantly lower and further aft for this case than for the higher Re = 3.6×10^6 case. This is due to the formation of a separation bubble at the surface of the cylinder, which occurs in a critical Reynolds number range [18].

Results for finite aspect ratio cylinders with two free ends in the Reynolds number range of the current experiments are more limited as most studies are concentrated on cantilevered finite aspect ratio geometries [22–24]. Zdravkovich et al. [25] investigated the pressure distributions around circular cylinders of finite aspect ratio with two free ends for the Reynolds number between 0.6 and 2.6×10^5. Data from that work for a L/D = 1 and Re = 2.6×10^5 showed $C_{p,min} \approx -1.6$ occurring at $\theta \approx 70°$. The angular location of $C_{p,min}$ was consistent with the data from Achenbach [18] with a difference of approximately 33% in $C_{p,min}$. The work in this study is compared with the results of Achenbach to provide comparative analysis on the effect of the short aspect ratio on the pressure distribution.

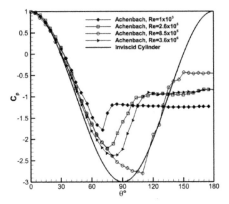

Figure 6. Pressure distribution around cylinder as a function of Reynolds number. Data from Achenbach [18].

3.2. Pressure Surveys

Figure 7 shows the surface pressure distribution along the centerline of the smooth cylinder for the Reynolds number range investigated. The data showed that $C_{p,min}$ decreased and its location moved aft as the Reynolds number increased, consistent with the trends for the 2D cylinder. At Re = 1.9 × 10^5 the surface pressure deviated almost immediately from the inviscid profile and had a minimum pressure coefficient of $C_{p,min} = -0.5$ at $\theta \approx 68°$. The pressure coefficient recovered slightly to a nominally constant value of $C_p = -0.4$ after this point. The location and value of $C_{p,min}$ were indicative of laminar flow around a cylinder. For the Re = 7.7 × 10^5 case the minimum pressure coefficient was found to be $C_{p,min} = -1.8$ at $\theta \approx 90°$, more consistent with turbulent flow. The dependence on the Reynolds number appeared to be more significant at lower Reynolds numbers and was likely a result of the transition from laminar to turbulent flow.

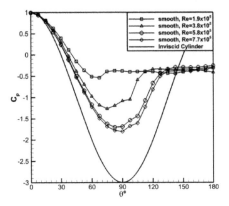

Figure 7. Surface pressure distribution for the smooth cylinder case.

The results for the valley of the center channel for the corrugated cylinder are shown in Figure 8. There was similar Reynolds number dependence observed for this model. The minimum pressure coefficient, $C_{p,min}$, decreased with the increased Reynolds number going from $C_{p,min} \approx -0.33$ at Re = 1.9 × 10^5 to $C_{p,min} \approx -1.24$ at Re = 7.7 × 10^5. While the location of $C_{p,min}$ shifted farther aft on the cylinder with increasing Reynolds number ($\theta \approx 54$ to $66°$), this shift was less significant than was observed for the smooth cylinder. In the separated wake of the corrugated cylinder, the pressure coefficient was nominally uniform for all cases at $C_p = -0.52$ except for the lowest Reynolds number case for which $C_p = -0.28$ in this region. This case also showed very little pressure recovery after the location of the minimum pressure.

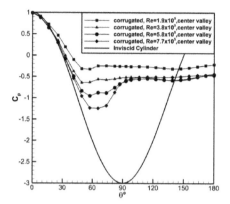

Figure 8. Pressure distributions through the corrugated cylinder center channel valley.

Comparison of the two current cases to previous data [18] are shown in Figure 9. The current data for the low aspect ratio smooth cylinder indicated that the pressure distribution was significantly altered by the lower aspect ratio of the model in the current study throughout the Reynolds number range investigated. Specifically, $C_{p,min}$ was lower and its location more aft for the infinite span compared to the finite span case. These observations indicated that end effects caused earlier separation, which reduced the pressure change on the cylinder surface. Data taken were also acquired with end plates (1.5D) on the finite aspect ratio smooth cylinder. These plates were insufficient to counteract end effects at low Re, as can be seen in Figure 9a; however, at the higher Re, shown in Figure 9b, the surface pressure profiles are more similar to the reference data of Achenbach [18]. The high Reynolds number cases for the current data and the reference data were both in the critical Reynolds number range with flow separation and reattachment; however, the gradient in the pressure distribution following separation was lower for the finite aspect ratio cylinder. This implied that the separation and reattachment region was likely smaller for the finite aspect ratio case in the critical Reynolds number range.

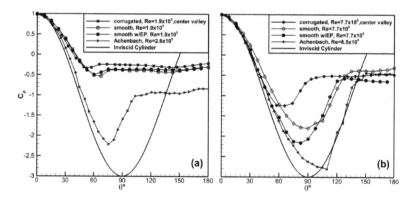

Figure 9. Comparison of the surface pressure distribution around the corrugated and smooth cylinders. (a) Re = 1.9×10^5, (b) Re = 7.7×10^5.

Comparison of the corrugated and smooth cylinders showed $C_{p,min}$ was significantly lower for the smooth cylinder when compared to the corrugated cylinder, as shown in Figure 9. The location of the measured minimum pressure and the minimum pressure coefficient, to the spatial resolution of the current data, are shown in Figure 10 to highlight the difference between the models. The location of the minimum pressure, θ_{min}, was earlier and the value of $C_{p,min}$ was higher for the corrugated cylinder versus the smooth cylinder. For example, at Re = 7.7×10^5 the values were -1.25 vs. -1.80 for $C_{p,min}$ and $57°$ vs. $90°$ for θ_{min}. The pressure surveys indicated that the separation was earlier for the corrugated model and was suggestive that the drag would be higher for case as well. Limited direct drag measurements, not shown in this work, measured a drag coefficient of $C_d = 0.95$ for the corrugated cylinder compared to $C_d = 0.60$ for the smooth cylinder at Re = 7.7×10^5 supporting the results of the pressure surveys. The higher relative $C_{p,min}$ values were suggestive that the flow speeds, and therefore the degree of flow acceleration, for the corrugated model were lower than for the smooth cylinder. This was important for the motivating accelerated wind application in that it indicated that the smooth cylinder would perform better. The trends in these results would be expected to hold for the real world application, which would include wind turbine blades, though the magnitudes would be expected to be different.

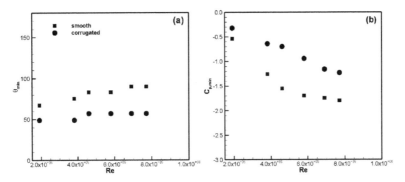

Figure 10. (a) θ_{min} and (b) $C_{p,min}$ versus Reynolds number for the smooth and corrugated cylinders.

The surface pressures at $\phi = 42°, 98°$ up the side wall of the corrugated cylinder model are shown in Figure 11. The Reynolds number trends observed in the channel valley continued along the channel walls with $C_{p,min}$ decreasing with increasing Reynolds number. At $\phi = 42°$ up the channel wall the location of $C_{p,min}$ moved from $\theta \approx 57°$ at Re = 1.9×10^5 to $\theta \approx 65°$ at Re = 3.8×10^5. Beyond this Reynolds number the separation point remained at nominally the same angular location. Similar trends were observed at $\phi = 98°$ up the side wall. It is interesting to note that stagnation conditions, $C_p < 1$ at $\theta = 0$, were not observed at the $\theta = 98°$ location indicating the leading edge flow field was quite complex and three dimensional.

The pressure distributions along the sidewalls are compared in Figure 12. The minimum pressure coefficient, $C_{p,min}$, increased from $C_{p,min} = -1.46$, in the valley, to $C_{p,min} = -1.23$ at $\phi = 42°$ and $C_{p,min} = -0.74$ at $\phi = 98°$. The angular location of $C_{p,min}$ was the same for the valley and $\phi = 42°$ but moved slightly further aft at the $\phi = 98°$. The pressure at the leading edge also showed a dependence on the wall location. In the channel valley, $C_p = 0.99$ indicating that the tap was at or near a stagnation point. The pressure coefficient was slightly less than this ($C_p = 0.97$) at $\phi = 42°$. In contrast, at $\phi = 98°$, $C_p = 0.85$ for the leading tap. The data show that the flow approached stagnation conditions in or near the valley bottom. The pressure coefficients were uniform in the wake region. The magnitudes of C_p show that the surface pressure was three dimensional within the channels. This is suggestive that the flow acceleration was most prominent in the valley region and decreased towards the side walls of the channel. Note that only a single Reynolds number is shown in Figure 12 for brevity; however, the qualitative trends were consistent for all Reynolds numbers investigated.

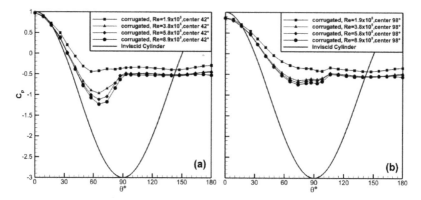

Figure 11. Pressure distributions at (a) $\phi = 42°$ and (b) $\phi = 98°$ up the channel wall.

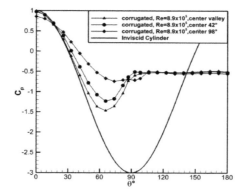

Figure 12. Pressure distribution as a function of channel wall location.

The corrugated cylinder had three channels along the axis of the model allowing for comparison of end versus interior channels. The pressure distribution in the valleys of the center (interior channel) and top channel (end channel) is shown in Figure 13. Significant differences between the interior and end channels were observed. The end channel had lower C_p values at all surface locations and for all Reynolds number cases indicating the end channels had higher potential for energy harvesting compared to interior channels. For example, $C_{p,min} = -1.86$ for the top channel, and $C_{p,min} = -1.46$ for the center channel at Re = 8.9×10^5. A similar trend was observed for the Re = 1.9×10^5 case with $C_{p,min}$ lower for the top channel (-0.55 vs. -0.32).

The difference between the two cases was caused by the channel boundary conditions, which were significantly different between the top and the middle channels. The interior channel had nominally symmetric boundary conditions due to the existence of channels above and below. The end channel had different boundary conditions on either side. The lower portion of the top channel was common with the upper portion of the center channel; however, the top portion of the upper channel was bounded by the free stream allowing air flow to go over top of the model. It was interesting to note that these end effects encountered by the top and bottom channels enhanced the performance of the flow dropping $C_{p,min}$ in the edge channels. The data suggested that the separation was delayed for the end channels resulting in the continued decrease in C_p, which for the intended application would be beneficial. Data with end plates, shown in Figure 13b, confirmed that end effects were responsible for the differences in the surface pressures in the center channel, though the end channel saw minimal improvement with the addition of the plates. The end plates did not appear to alter the pressure on the downstream side of the corrugated cylinder indicating that drag would likely continue to be high for this model.

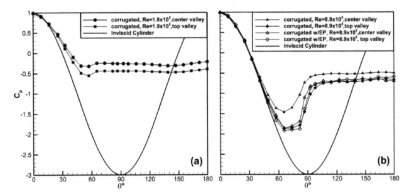

Figure 13. Comparison of pressure distribution in the center and top channels. (a) Re = 1.9×10^5, (b) Re = 7.7×10^5.

3.3. Velocity Surveys

Velocity surveys were conducted with single-wire hot-wire probes around the smooth and corrugated cylinders. The probe was oriented at each angular location such that it measured the azimuthal velocity component, $V_\theta(r)$. The azimuthal velocity was of primary interest in this study as it is the velocity component normal to the intended rotor plane. Velocity surveys were conducted normal to the surface (i.e., in the radial direction) at 15° increments around the cylinder starting at the leading edge, $\theta = 0°$. The data for the smooth cylinder were acquired at the mid-height of the cylinder, while the data for the corrugated cylinder were acquired along the center channel valley of the model. We note that both measurement locations were expected to have small relative spanwise velocity components due to the presence of the corrugations and/or symmetry. The radial velocity component was also expected to be small outside of the model wakes, which were beyond the expected placement of the rotors.

Results of the velocity surveys are shown in contour form in Figure 14 for the Re = 7.7×10^5 case, which was closest to the expected operational Reynolds number. The dash-dot-dot line in the figure marks the boarder where the flow speed was equal to the free stream. This was included to differentiate the regions where the flow speed was either above or below the free stream value. Both cases show deceleration as the flow approached the leading edge, as expected. The flow then accelerated around both the smooth and corrugated cylinders; however, the velocity fields were quantitatively different for the two models. The data clearly show that the location of the peak flow speeds was shifted to higher angles (i.e., closer to $\theta \approx 90°$) for the smooth cylinder. The peak speed for the smooth cylinder occurred at $\theta \approx 90°$ after which the flow decelerated, as shown in Figure 14a. In contrast, the peak flow speed occurred earlier at $\theta \approx 75°$ for the corrugated cylinder model. The results also showed that the wake region, demarked by the region where the flow speed was below the free stream speed for the aft portion of the model, was physically larger for the corrugated cylinder. These observations were consistent with the pressure surveys, which showed lower C_p values and more pressure recovery in the wake for the smooth cylinder. They are also consistent with the lower drag measurements for the smooth cylinder. It is noted that single hot-wire probes are not able to resolve the direction of the flow, only the magnitude, in the wake. As a result, the detailed structure of the wake region, e.g., where the flow could be reversed, cannot be determined from the current data.

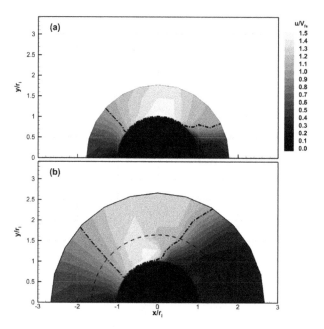

Figure 14. Velocity distribution around the (**a**) smooth and (**b**) and corrugated cylinders. Note: Solid black denotes minor diameter of the corrugated cylinder and the diameter of the smooth cylinder. The dashed line in (**b**) represents the location of the major diameter. The dash-dot-dash line shows the contour of $u/V_{fs} = 1$. Re = 7.7×10^5. Flow left to right.

Velocity profiles at select angles are shown in Figure 15 to quantify the differences between the velocity fields for the two cases. The radial distance was rescaled in Figure 15 so that the model surface occurred at $r^* = 0$, and the distance was normalized by the difference in height between the major and minor axis of the corrugated cylinder. This scaling method resulted in $r^* = 1$ corresponding to the edge of the major diameter in the corrugated model and the equivalent dimensional distance for the smooth cylinder. The inviscid velocity profile at $\theta = 90°$ (dash-dot) and free stream speed (dashed) are also plotted for reference in Figure 15.

The velocity profiles were found to be qualitatively similar for the two geometries at $\theta = 75°$. At this angular location the velocities were higher than the freestream at all radial measurement locations, with the highest value of $V_\theta/V_{fs} \approx 1.43$ at the measurement location closest to the surface. The velocity decreased to $V_\theta/V_{fs} \approx 1.2$ at the upper measurement location and appeared to be asymptotically approaching the freestream value. The boundary layer at the cylinder surface was relatively thin with a thickness less than the distance from the wall to the first measurement point at $r^* = 0.06$.

At 90° the velocity profiles were qualitatively different for the two geometries. The flow around the smooth cylinder continued to accelerate due to the curvature of the wall. Velocities were higher at all radial locations compared to the 75° location for this geometry, and the boundary layer remained comparatively thin. The maximum value of the angular velocity was $V_\theta/V_{fs} \approx 1.6$, which was approximately 20% lower than the inviscid velocity of $V_\theta/V_{fs} \approx 2.0$. The corrugated cylinder showed a qualitatively different profile at this angular location. The changes in the structure of the velocity profile were most pronounced near the wall. The location of the peak velocity moved away from the wall forming what appeared to be a thick viscous boundary layer that occupied approximately 20% of the channel height. This resulted in a noticeably less uniform velocity profile, particularly near the wall. Additionally, the flow speed was lower than the freestream for approximately the bottom (i.e., near wall) 10% of the channel height. The flow speeds were comparatively lower at all locations than were measured at $\theta = 75°$ indicating the flow was decelerating within the channel.

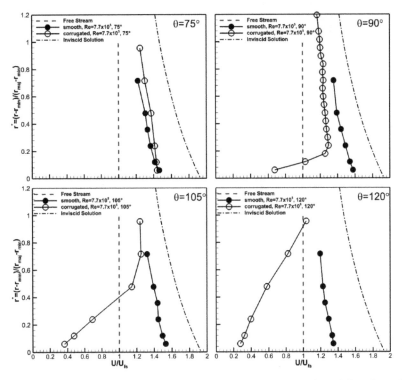

Figure 15. V_θ velocity profiles normal to the surface. Angular position as indicated. Re = 7.7×10^5.

These trends in the velocity data continued at $\theta = 105°$. Here the velocity profile for the smooth cylinder was nominally the same as it was at $\theta = 90°$, though the velocity magnitudes were slightly reduced. Comparison showed a slight deceleration in the velocity for the entire profile; however, this was expected due to the change in curvature of the model. The viscous boundary layer remained thinner than the data spacing. The low speed region of the corrugated cylinder continued to expand out away from the valley wall filling the bottom 60% of the channel height. Flow speeds lower than the free stream were measured in the bottom 40% of the channel height. By 120° the velocity magnitude was below the free stream in 80% of the corrugated channel height. The velocities also continued to decrease for the smooth cylinder, though they were still above the freestream value at all measurement locations. The velocity data were suggestive that viscous effects, likely due to the increased surface area of the corrugated cylinder, were responsible for the differences in the velocity and pressure results observed for that model.

Recall that the purpose of placing wind turbine blades next to a surface is to increase the kinetic energy of the wind before it enters the wind turbines blades. Ideally, one would prefer the flow speeds to be as high as possible (to maximize energy harvest) and uniform across the blades (for structural reasons). The experimentally measured and the computed inviscid velocity profiles were used to provide an estimate of the kinetic energy flux per unit width by:

$$KE = \int_0^1 \frac{V_\theta^3(r^*)}{U_{FS}^3} dr^* \qquad (3)$$

It is acknowledged that the flow field varies in two dimensions (i.e., in the r-z plane) at a particular θ location for the two experimental geometries and that the actual kinetic energy flux over an area would differ from the integrated values using Equation (3). However, the estimate from this 1D integration provided a quantitative comparison of the geometries with the current data.

The inviscid profile provides a maximum of 5 times the kinetic energy in the free stream at $\theta = 90°$, as shown in Figure 16. This is less than the "8 times" value because the inviscid velocity distribution varies with r as previously discussed. The smooth cylinder also showed a peak normalized kinetic energy at $\theta = 90°$, though it was lower than for the inviscid flow (2.8 vs. 5). This result highlights the importance of the cubic functionality of the power with velocity. The corrugated cylinder had a maximum normalized kinetic energy at $\theta = 75°$ and was comparable in magnitude (~2.5) to the smooth cylinder at this location. The normalized kinetic energy decreased for the corrugated cylinder at $\theta = 90°$ and was lower than for the smooth cylinder case (1.7 vs. 2.8) and less than half of the inviscid case.

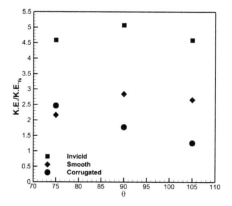

Figure 16. Normalized kinetic energy from measured velocity profiles. Re = 7.7×10^5.

It is instructive to note that while both model cases were significantly below the inviscid kinetic energy potential, they were both above that for a rotor alone in a free stream (i.e., a value of 1). This indicated that the rotors could potentially experience enhancement due to placement next to both models. Interpretation of the values for the corrugated cylinder should be done carefully as the channel pressure and velocity data were suggestive that the velocities near the bounding walls would be lower than in the centerline of the channel. One should therefore expect that the kinetic energy flux (i.e., the area integrated KE flux) in the corrugated cylinder would be lower than what was calculated using Equation (3).

The turbine blades for the real-world application are best placed at $\theta = 90°$ for operational reasons. These reasons include the ability to self-align with the changes in the wind direction due to forces on the symmetrically placed pairs of rotor blades and the need for the system to be aligned out of the wind if needed. This azimuthal position therefore deserves additional discussion. Comparison of the data at $\theta = 90°$ clearly shows that the smooth cylinder had both higher velocities and a more uniform velocity profile. Uniformity in the velocity profile is desirable for blade loading and structural reasons in an accelerated application. In particular, the variation in the velocity profile measured for the corrugated model near the wall, approximately 50%, represents a potential difficulty if used as the loading near the tip varies significantly. The smooth cylinder on the other hand experiences a smaller fractional change, approximately 12%, across the blades. The lower velocities for the corrugated cylinder resulted in a 38% drop in the kinetic energy potential for the corrugated cylinder at this angular location. The results clearly showed that the smooth cylinder was a more desirable platform shape for the larger project.

4. Discussion

The pressure and velocity fields around short aspect ratio (AR = 0.93) smooth and corrugated cylinders were investigated experimentally using surface pressure and single-wire hot-wire anemometry. The corrugated cylinder was originally conceived as a strategy to accelerate and direct the wind into wind turbine blades to provide a scalable midrange "accelerated wind" system, which motivated this study. The smooth cylinder provided a canonical reference geometry for comparative purposes as well as a second candidate platform design. The smooth cylinder results are unique in that they were conducted for a smaller aspect ratio than is typically studied and for a non-surface mounted finite aspect ratio model. The goals of these experiments were to quantify and compare the surface pressures and flow fields around the low aspect ratio (AR = 0.93) smooth and corrugated cylinders. The experiments were performed without wind turbine blades, which represents the upper limiting condition for the motivating accelerated wind application. This then provides a "best case" scenario by which the designs could be compared.

The results indicated that the minimum pressure coefficient, $C_{p,min}$, was higher and the flow separated earlier for the smooth finite aspect ratio cylinder when compared to published results for a bounded (i.e., "infinite aspect ratio") cylinder at similar Reynolds numbers. These results showed that the end effects on the finite aspect ratio cylinder played a significant role even at the center of the small aspect ratio cylinder and were consistent with prior work [18,25], which was conducted near the Re of the current work. These results then indicate that end effects are of critical importance when using low aspect ratio cylindrical bodies to accelerate the flow for wind turbine applications.

The corrugated cylinder showed smaller decreases in the pressure coefficient, C_p, and earlier separation compared to the smooth cylinder with the same aspect ratio. This occurred over the entire Reynolds number range investigated. Both of these results indicated that flow acceleration around the corrugated cylinder was lower than for the smooth cylinder. The early separation also resulted in an increased the size of the downstream wake and subsequently an increase in the drag coefficient for the corrugated cylinder. The azimuthal velocity measurements, V_θ, confirmed that the flow speeds around the corrugated cylinder were reduced when compared to the smooth cylinder. The velocity profiles revealed that the cause of the differences between the models was the development of a large viscous region in the channels due to the presence of the corrugation walls. There was a measurable variation in the pressure fields between end and internal channels for the corrugated cylinder indicating that end effects were also important for this geometry. Interestingly, the edge channels appeared to perform better (based on magnitude and location of $C_{p,min}$) than the center channel did. The end effects were mitigated by placing bounding plates on the upper and lower surfaces of the model, reducing channel to channel variation.

Both the smooth and corrugated cylinder models had azimuthal velocities that were below those for inviscid flow around a cylinder as expected given the viscous nature of real flows. However, both cases did show increased flow speeds compared to the free stream in the intended rotor plane. This resulted in a best case of 58% kinetic energy harvesting capability for the smooth cylinder case and 34% for the corrugated cylinder when compared to the idealized upper limit based on the inviscid flow case. This result provides useful insight for other accelerated wind applications, e.g., building augmented wind turbines, where turbines are located near structures that were not specifically designed to provide flow acceleration. In these cases, the results of this work show that shapes that may be suboptimal from an aerodynamic standpoint may still result in accelerated flow (compared to the free stream wind speed), which could be utilized for energy harvesting.

The following conclusions for the specific "accelerated wind" goal of this project were made based on these results. End effects must be considered in platforms designed for accelerated wind applications as they influence the velocity and pressure distributions in a negative manner. These can be mitigated when designing the shape of the end of the platforms. The earlier separation and larger wake region experienced by the corrugated model was consequential for design of a real world platform as it resulted in higher drag. This would require higher structural requirements if the corrugated model were to be chosen. Consider next the velocity entering the location of the wind turbine blade placement ($\theta = 90°$). Kinetic energy considerations aside, a uniform velocity profile is desirable from a turbine blade structural standpoint. The azimuthal velocity profile in the planned rotor plane was significantly more uniform when compared to that measured for the corrugated cylinder indicating again that the smooth cylinder was a more favorable model moving forward.

The smooth cylinder was found to outperform the corrugated cylinder based on all of the study test metrics: magnitude and location of $C_{p,min}$, magnitude and uniformity of V_θ in the intended rotor plane, and the line integrated kinetic energy. From this it was concluded that the smooth cylinder clearly had higher potential for application in designed accelerated wind applications. The results of this work directly impacted the higher level goals of the project. Specifically, the smooth cylinder model was selected, while the corrugated model was abandoned, in the continuing design process due to the measurably poor performance of the corrugated model in this study.

Author Contributions: The data presented in this paper were the work of M.P. and D.B. M.P. was the graduate student who was responsible for the data acquisition, processing and analysis detailed in this work. D.B. was the primary investigator of this work and graduate advisor for M.P. All authors have read and agreed to the published version of the manuscript.

Funding: This work was funded through the New York State Energy and Research Development Authority Grant# 10596.

Conflicts of Interest: The authors declare no conflict of interest. Optiwind Inc. provided the specifications for the corrugated cylinder used in this study. Optiwind had no role in the study design, performance of the research, analysis of the data or writing and publication of this work.

References

1. Johnson, G.L. *Wind Energy Systems*; Prentice-Hall: Englewood Cliffs, NJ, USA, 1985.
2. Cheng, J.H.; Hu, S.Y. Innovatory designs for ducted wind turbines. *Renew. Energy* **2008**, *33*, 1491–1498.
3. Anzai, A.; Nemoto, Y.; Ushiyama, I. Wind tunnel analysis of concentrators for augmented wind turbines. *Wind Eng.* **2004**, *28*, 605–614. [CrossRef]
4. Van Bussel, G.J.W. The science of making more torque from wind: Diffuser experiments and theory revisited. *J. Phys. Conf. Ser.* **2007**, *75*, 12010. [CrossRef]
5. Ohya, Y.; Karasudani, T.; Sakurai, A.; Abe, K.I.; Inoue, M. Development of a shrouded wind turbine with a flanged diffuser. *J. Wind Eng. Ind. Aerod.* **2008**, *96*, 524–539. [CrossRef]
6. Gilbert, B.L.; Foreman, K.M. Experimental Demonstration of a Diffuser-Augmented Wind Turbine Concept. *J. Energy* **1979**, *3*, 235–240. [CrossRef]
7. Phillips, D.G.; Nash, T.A.; Oakey, A.; Flay, R.G.J.; Richards, P.J. Computational fluid dynamic and wind tunnel modelling of a Diffuser Augmented Wind Turbine. *Wind Eng.* **1999**, *23*, 7–13.
8. Bagheri-Sadeghi, N.; Helenbrook, B.T.; Visser, K.D. Ducted wind turbine optimization and sensitivity to rotor position. *Wind Energy Sci.* **2018**, *3*, 221–229. [CrossRef]
9. Kanya, B.; Visser, K.D. Experimental validation of a ducted wind turbine design strategy. *Wind Energy Sci.* **2018**, *3*, 919–928. [CrossRef]
10. Venters, R.; Helenbrook, B.T.; Visser, K.D. Ducted Wind Turbine Optimization. *J. Sol. Energy Eng.* **2018**, *140*. [CrossRef]
11. Bahadori, M. Wind Tower Augmentation of Wind Turbines. *Wind Eng.* **1984**, *18*, 144–151.
12. Duffy, R.E.; Jaran, C.; Ungermann, C. *Aerodynamic Characteristics of the TARP (Toroidal Accelerator Rotor Platform) Wind Energy Conversion System*; Rensselaer Polytechnic Inst.: Troy, NY, USA, 1980.
13. Duffy, R.E.; Liebowitz, B. *Verification Analysis of the Performance of the Toroidal Accelerator Rotor Platform Wind Energy Conversion System*; NYSERDA: Albany, NY, USA, 1988.
14. Weisbrich, A.L.; Ostrow, S.L.; Padalino, J.P. WARP: A Modular Wind Power System for Distributed Electric Utility Application. *IEEE Trans. Ind. Appl.* **1996**, *32*, 778–787. [CrossRef]
15. Bohl, D.; Helenbrook, B.; Kanya, B.; Visser, K.; Marvin, R.; Mascarenhas, B.; Parker, M.; Rocky, D. Analysis and Design of a Wind Turbine with a Wind Accelerator. In Proceedings of the 29th AIAA Applied Aerodynamics Conference, Honolulu, HA, USA, 27–30 June 2011.

16. Ameri, A.A.; Rashidi, M. Analysis of a Concept for a Low Wind Speed Tolerant Axial Wind Turbine. In Proceedings of the ASME Turbo Expo: Turbine Technical Conference and Exposition, Vancouver, BC, Canada, 6–10 June 2011.
17. Prandtl, L.; Tiejens, O.G. *Fundamentals of Hydro- and Aeromechanics*; Dover Publications: New York, NY, USA, 1934.
18. Achenbach, E. Distribution of local pressure and skin friction around a circular cylinder in cross-flow up to Re = 5 × 10^6. *J. Fluid Mech.* **1968**, *34*, 625–639. [CrossRef]
19. Achenbach, E. Influence of surface roughness on the cross-flow around a circular cylinder. *J. Fluid Mech.* **1971**, *46*, 321–335. [CrossRef]
20. Roshko, A. *On the Drag and Shedding Frequency of Two-Dimensional Bluff Bodies*; National Adivsory Committee for Aeronautics: Washington, DC, USA, 1954.
21. Buranall, W.; Loftin, L. *Experimental Investigation of the Pressure Distribution about a Yawed Circular Cylinder in the Critical Reynolds Number Range*; National Advisory Committee for Aeronautics, Ed.; National Advisory Committee for Aeronautics: Washington, WA, USA, 1951.
22. Park, C.W.; Lee, S.J. Flow structure around a finite circular cylinder embedded in various atmospheric boundary layers. *Fluid Dyn. Res.* **2002**, *30*, 197–215. [CrossRef]
23. Park, C.W.; Lee, S.J. Free end effects on the near wake flow structure behind a finite circular cylinder. *J. Wind Eng. Ind. Aerod.* **2000**, *88*, 231–246. [CrossRef]
24. Uematsu, Y.; Yamada, M.; Ishii, K. Some Effects of Free-Stream Turbulence on the Flow Past a Cantilevered Circular-Cylinder. *J. Wind Eng. Ind. Aerod.* **1990**, *33*, 43–52. [CrossRef]
25. Zdravkovich, M.M.; Brand, V.P.; Mathew, G.; Weston, A. Flow Past Short Circular-Cylinders with 2 Free Ends. *J. Fluid Mech.* **1989**, *203*, 557–575. [CrossRef]

© 2020 by the authors. Licensee MDPI, Basel, Switzerland. This article is an open access article distributed under the terms and conditions of the Creative Commons Attribution (CC BY) license (http://creativecommons.org/licenses/by/4.0/).

Article

Derivation of the Adjoint Drift Flux Equations for Multiphase Flow

Shenan Grossberg [1], Daniel S. Jarman [2] and Gavin R. Tabor [1,*]

1. College of Engineering, Maths and Physical Sciences, University of Exeter, Harrison Building, North Park Road, Exeter EX4 4QF, UK; s.grossberg@exeter.ac.uk
2. Hydro International Ltd., Shearwater House, Clevedon Hall Estate, Victoria Road, Clevedon BS21 7RD, UK; djarman@hydro-int.com
* Correspondence: g.r.tabor@exeter.ac.uk

Received: 24 January 2020; Accepted: 07 March 2020; Published: 11 March 2020

Abstract: The continuous adjoint approach is a technique for calculating the sensitivity of a flow to changes in input parameters, most commonly changes of geometry. Here we present for the first time the mathematical derivation of the adjoint system for multiphase flow modeled by the commonly used drift flux equations, together with the adjoint boundary conditions necessary to solve a generic multiphase flow problem. The objective function is defined for such a system, and specific examples derived for commonly used settling velocity formulations such as the Takacs and Dahl models. We also discuss the use of these equations for a complete optimisation process.

Keywords: adjoint optimization; multiphase flow; computational fluid dynamics

1. Introduction

The adjoint method is currently attracting significant interest as an optimization process in CFD. The objective of the adjoint approach is to calculate the sensitivity of the flow solution with respect to changes in the input parameters, most commonly changes in the geometry. This can then in principle be used as the basis for an iterative optimization algorithm based on gradient information (the sensitivities) which can optimize the design with many fewer function evaluations than would be the case for non-gradient-based approaches (such as genetic algorithms). Calculating the sensitivities requires differentiating the governing equations with respect to the changes of the input parameters, and since the governing equations for fluid flow are the Navier-Stokes equations (or equations derived from these such as the Reynolds Averaged Navier Stokes equations), this is understandably very challenging. There are two main approaches; the discrete adjoint approach, and the continuous adjoint approach. In the discrete adjoint approach, the sensitivity matrix is calculated numerically by evaluating the system for small changes in the inputs and applying standard finite difference methods. In the continuous adjoint approach, the sensitivities are calculated mathematically using lagrange multipliers. This is more elegant and provides an implementation which is easier to code, requires fewer evaluations and can be made numerically consistent with the evaluation of the original equation set. However it does require significant mathematical analysis in advance, and if the problem formulation changes (different equations, boundary conditions etc) this has to be repeated. Examples of the application of the continuous adjoint method for single phase flow can be found in a range of areas [1,2] such as automotive [3–5], aerospace [6,7] and turbomachinery [8–10], and implementations of the equations can be found in general purpose CFD codes such as STAR-CCM, ANSYS Fluent [11] and Engys Helyx [4]. However the equations are complex to develop and application to multiphase systems is only just starting [12]. In many cases, even just the evaluation of the sensitivities is valuable, as they can be used to indicate possible changes to the design engineers. Beyond this the sensitivities can also be used as the basis for an optimization loop [2]. This of course necessitates the morphing of the geometry through techniques such as volumetric B-splines [13] or Radial Basis Functions [14] and consequent updating of the mesh [15].

Multiphase flow is the simultaneous flow of two or more immiscible phases in a system. In dispersed multiphase systems, one or more of the phases exists as fluid particles small enough not to be resolved in the simulation; examples include gas bubbles in water, emulsions (liquid droplets in another immiscible liquid) and actual solid particles in gas or liquid. A wide variety of different mathematical models have been derived over the years to describe dispersed multiphase flow, including mixture models, lagrangian particle tracking, and eulerian n-fluid models [16,17]. Which is used depends on the exact physics of the problem, as well as factors such as available computing resources and desired accuracy. In many physical systems, the density ratio between the two phases is low, generally less than 2:1, and the drag force between them is high. Therefore, to a good approximation, the two phases can be considered to respond to pressure gradients as a single phase. Additionally, the slip (drift) between the phases is primarily due to the gravitational settling of the dispersed phase. This might adequately describe solid particles in water or an emulsion of immiscible liquids, and in these cases a commonly used mathematical model is the drift flux model. Hence it is this set of equations we have decided to focus on.

In the drift flux model, the two phases are treated as one: the momentum and continuity equations for both phases are summed to create a mixture-momentum and mixture-continuity equation, and the transport of the dispersed phase is modelled using a drift equation. The three equations, collectively called the drift flux equations, are listed below:

$$\frac{\partial \rho_m \mathbf{v}_m}{\partial t} + (\mathbf{v}_m \cdot \nabla)(\rho_m \mathbf{v}_m) = -\nabla(\rho_m p_m) + \nabla \cdot (2\mu_m D(\mathbf{v}_m))$$
$$- \nabla \cdot \left(\frac{\alpha}{1-\alpha} \frac{\rho_c \rho_d}{\rho_m} \mathbf{v}_{dj} \mathbf{v}_{dj} \right) + \rho_m \mathbf{g} + F, \qquad (1a)$$

$$\frac{\partial \rho_m}{\partial t} + \nabla \cdot (\rho_m \mathbf{v}_m) = 0, \qquad (1b)$$

$$\frac{\partial \alpha}{\partial t} + \nabla \cdot (\alpha \mathbf{v}_m) = -\nabla \cdot \left(\frac{\alpha \rho_c}{\rho_m} \mathbf{v}_{dj} \right) + \nabla \cdot K \nabla \alpha, \qquad (1c)$$

where:

- α is the dispersed-phase volume fraction,
- ρ_c is the continuum density,
- ρ_d is the dispersed-phase density,
- ρ_m is the mixture density, defined as $\alpha \rho_d + (1-\alpha)\rho_c$,
- $\mathbf{v}_m$ is the mixture velocity,
- p_m is the mixture kinematic pressure,
- μ_m is the mixture viscosity, defined as the sum of the continuum, dispersed-phase and mixture turbulent viscosities, $\mu_c + \mu_d + \mu_m^t$,
- $D(\mathbf{v}_m) = \frac{1}{2} \left(\nabla \mathbf{v}_m + (\nabla \mathbf{v}_m)^T \right)$ is the mixture strain rate tensor,
- $\mathbf{v}_{dj}$ is the dispersed-phase settling velocity,
- $\mathbf{g}$ is the acceleration due to gravity,
- F is the capillary force and
- K is the turbulent diffusion coefficient, defined as the mixture eddy diffusivity, $\nu_m^t = \frac{\mu_m^t}{\rho_m}$.

In summing the momentum equations, not only have the number of equations been reduced from four to three, but the inter-phase momentum transfer terms have also been eliminated which were numerically unstable [18]. Hence, a far more robust equation set has been produced and the computational resources required to solve the system have been reduced. This also makes it a very appropriate basis from which to develop an adjoint formulation suitable for applying to dispersed multiphase flows in this regime. This is the challenge of the current paper. We focus in particular on wall-bounded or ducted flows, in which there is no contribution to the objective function from the interior of the domain, in other words, the performance of the system is entirely governed by the boundary properties.

The paper is organized as follows. The optimization problem is stated in Section 2 and the adjoint equations for the drift flux model are derived for the general case in Section 2. These equations are then applied to the specific case of ducted or wall-bounded flows in Section 3, with the objective function for this case being specified in Section 4, and different settling velocities in Section 5. Finally, the conclusions follow in Section 6.

2. The Optimization Problem

If the performance of a device is measured by an objective function, J, and the residuals of the primal (flow) equations are given by $\mathcal{R}$, the optimisation problem can be stated as,

$$\text{optimise } J(x,y) \text{ subject to } \mathcal{R}(x,y) = 0, \tag{2}$$

where x are the design parameters and y are the primal variables [19]. It can then be formulated as,

$$\mathcal{L} = J + \int_\Omega \lambda \mathcal{R} \, d\Omega, \tag{3}$$

where $\mathcal{L}$ is the Lagrange function, λ are the Lagrange multipliers (also referred to as the adjoint variables) and Ω is the flow domain. In this case, the primal equations are the steady state drift flux equations, with the capillary force taken to be zero [18] and a Darcy term included in the mixture-momentum equation. They are rearranged in terms of their residuals, $\mathcal{R} = (R_1, R_2, R_3, R_4, R_5)^T$, as follows:

$$(R_1, R_2, R_3)^T = (\mathbf{v}_m \cdot \nabla)(\rho_m \mathbf{v}_m) + \nabla(\rho_m p_m) - \nabla \cdot (2\mu_m D(\mathbf{v}_m))$$
$$+ \nabla \cdot \left(\frac{\alpha}{1-\alpha} \frac{\rho_c \rho_d}{\rho_m} \mathbf{v}_{dj} \mathbf{v}_{dj} \right) - \rho_m \mathbf{g} + \aleph \rho_m \mathbf{v}_m, \tag{4a}$$

$$R_4 = -\nabla \cdot (\rho_m \mathbf{v}_m), \tag{4b}$$

$$R_5 = \nabla \cdot (\alpha \mathbf{v}_m) + \nabla \cdot \left(\frac{\alpha \rho_c}{\rho_m} \mathbf{v}_{dj} \right) - \nabla \cdot K \nabla \alpha, \tag{4c}$$

where $\aleph$ is the porosity, associated with the Darcy term. The variation of the Lagrange function with respect to the primal variables, $(\mathbf{v}_m, p_m, \alpha)$, and the design parameter, $\aleph$, is,

$$\delta \mathcal{L} = \delta_{\mathbf{v}_m} \mathcal{L} + \delta_{p_m} \mathcal{L} + \delta_\alpha \mathcal{L} + \delta_\aleph \mathcal{L}, \tag{5}$$

where, for example, $\delta_\alpha \mathcal{L} = \mathcal{L}(\alpha + \delta\alpha) - \mathcal{L}(\alpha)$. We choose the adjoint variables, $(\mathbf{u}, q, \beta) = (u_1, u_2, u_3, q, \beta)$, so that the variation with respect to the primal variables vanishes, i.e.,

$$\delta_{\mathbf{v}_m} \mathcal{L} + \delta_{p_m} \mathcal{L} + \delta_\alpha \mathcal{L} = 0, \tag{6}$$

and the Lagrange function now varies only with respect to the design parameter,

$$\delta \mathcal{L} = \delta_\aleph \mathcal{L} = \delta_\aleph J + \int_\Omega (\mathbf{u}, q, \beta) \delta_\aleph \mathcal{R} \, d\Omega. \tag{7}$$

Derivation of the Adjoint Drift Flux Equations

The adjoint drift flux equations are derived by substituting Equation (3) into Equation (6), giving,

$$\delta_{\mathbf{v}_m} J + \delta_{p_m} J + \delta_\alpha J$$
$$+ \int_\Omega (\mathbf{u}, q, \beta) \delta_{\mathbf{v}_m} \mathcal{R} \, d\Omega + \int_\Omega (\mathbf{u}, q, \beta) \delta_{p_m} \mathcal{R} \, d\Omega + \int_\Omega (\mathbf{u}, q, \beta) \delta_\alpha \mathcal{R} \, d\Omega = 0, \quad (8)$$

which can be expanded to,

$$\delta_{\mathbf{v}_m} J + \delta_{p_m} J + \delta_\alpha J + \int_\Omega d\Omega \, \mathbf{u} \cdot \delta_{\mathbf{v}_m}(R_1, R_2, R_3)^T + \int_\Omega d\Omega \, q \delta_{\mathbf{v}_m} R_4 + \int_\Omega d\Omega \, \beta \delta_{\mathbf{v}_m} R_5$$
$$+ \int_\Omega d\Omega \, \mathbf{u} \cdot \delta_{p_m}(R_1, R_2, R_3)^T + \int_\Omega d\Omega \, q \delta_{p_m} R_4 + \int_\Omega d\Omega \, \beta \delta_{p_m} R_5$$
$$+ \int_\Omega d\Omega \, \mathbf{u} \cdot \delta_\alpha(R_1, R_2, R_3)^T + \int_\Omega d\Omega \, q \delta_\alpha R_4 + \int_\Omega d\Omega \, \beta \delta_\alpha R_5 = 0. \quad (9)$$

The variation of $\mathcal{R}$ with respect to the primal variables can be determined as:

$$\delta_{\mathbf{v}_m}(R_1, R_2, R_3)^T = (\delta \mathbf{v}_m \cdot \nabla)(\rho_m \mathbf{v}_m) + (\mathbf{v}_m \cdot \nabla)(\rho_m \delta \mathbf{v}_m) - \nabla \cdot (2\mu_m D(\delta \mathbf{v}_m))$$
$$- \nabla \cdot (2\delta_{\mathbf{v}_m} \mu_d D(\mathbf{v}_m)) + \aleph \rho_m \delta \mathbf{v}_m, \quad (10a)$$

$$\delta_{\mathbf{v}_m} R_4 = -\nabla \cdot (\rho_m \delta \mathbf{v}_m), \quad (10b)$$

$$\delta_{\mathbf{v}_m} R_5 = \nabla \cdot (\alpha \delta \mathbf{v}_m), \quad (10c)$$

$$\delta_{p_m}(R_1, R_2, R_3)^T = \nabla(\rho_m \delta p_m), \quad (10d)$$

$$\delta_{p_m} R_4 = 0, \quad (10e)$$

$$\delta_{p_m} R_5 = 0, \quad (10f)$$

$$\delta_\alpha (R_1, R_2, R_3)^T = (\rho_d - \rho_c)\big((\mathbf{v}_m \cdot \nabla)(\delta \alpha \mathbf{v}_m) + \nabla(\delta \alpha p_m) + \delta \alpha (\aleph \mathbf{v}_m - \mathbf{g})\big)$$
$$- \nabla \cdot (2\delta_\alpha \mu_d D(\mathbf{v}_m)) + \nabla \cdot \delta_\alpha (\alpha \rho_d \mathbf{v}_{dj} \mathbf{v}_{dj}), \quad (10g)$$

$$\delta_\alpha R_4 = -(\rho_d - \rho_c) \nabla \cdot (\delta \alpha \mathbf{v}_m), \quad (10h)$$

$$\delta_\alpha R_5 = \nabla \cdot (\delta \alpha \mathbf{v}_m) + \nabla \cdot \delta_\alpha (\alpha \mathbf{v}_{dj}) - \nabla \cdot \left(\frac{\mu_m^t}{\rho_c} \nabla \delta \alpha\right)$$
$$+ \frac{\mu_m^t}{\rho_c} \left(\frac{\rho_d}{\rho_c} - 1\right) \nabla \cdot (\delta \alpha \nabla \alpha) + \frac{\mu_m^t}{\rho_c} \left(\frac{\rho_d}{\rho_c} - 1\right) \nabla \cdot (\alpha \nabla \delta \alpha). \quad (10i)$$

Derivation of Equations (10a), (10g) and (10i) can be found in Appendices A–C, respectively, where the variation of μ_m^t has been neglected. This is correct only for laminar flow regimes. For turbulent flows, neglecting this variation constitutes a common approximation, known as *frozen turbulence* [19]. This may introduce errors into the optimisation [20], although there are cases in the literature where the frozen turbulence assumption can be demonstrated to be acceptable [21].

With these variations, Equation (9) now reads,

$$\delta_{v_m} J + \delta_{p_m} J + \delta_\alpha J$$
$$+ \int_\Omega d\Omega\, \mathbf{u} \cdot \Big((\delta \mathbf{v}_m \cdot \nabla)(\rho_m \mathbf{v}_m) + (\mathbf{v}_m \cdot \nabla)(\rho_m \delta \mathbf{v}_m) - \nabla \cdot (2\mu_m D(\delta \mathbf{v}_m)) $$
$$- \nabla \cdot (2\delta_{v_m} \mu_d D(\mathbf{v}_m)) + \aleph \rho_m \delta \mathbf{v}_m \Big) - \int_\Omega d\Omega\, q \nabla \cdot (\rho_m \delta \mathbf{v}_m) + \int_\Omega d\Omega\, \beta \nabla \cdot (\alpha \delta \mathbf{v}_m)$$
$$+ \int_\Omega d\Omega\, \mathbf{u} \cdot \nabla(\rho_m \delta p_m) + (\rho_d - \rho_c) \int_\Omega d\Omega\, \mathbf{u} \cdot \big((\mathbf{v}_m \cdot \nabla)(\delta \alpha \mathbf{v}_m) + \nabla(\delta \alpha p_m) + \delta \alpha (\aleph \mathbf{v}_m - \mathbf{g}) \big)$$
$$- \int_\Omega d\Omega\, \mathbf{u} \cdot \Big(\nabla \cdot (2\delta_\alpha \mu_d D(\mathbf{v}_m)) \Big) + \int_\Omega d\Omega\, \mathbf{u} \cdot \Big(\nabla \cdot \delta_\alpha (\alpha \rho_d \mathbf{v}_{dj} \mathbf{v}_{dj}) \Big)$$
$$- (\rho_d - \rho_c) \int_\Omega d\Omega\, q \nabla \cdot (\delta \alpha \mathbf{v}_m) + \int_\Omega d\Omega\, \beta \Big(\nabla \cdot (\delta \alpha \mathbf{v}_m) + \nabla \cdot \delta_\alpha (\alpha \mathbf{v}_{dj}) - \nabla \cdot \Big(\frac{\mu_m^t}{\rho_c} \nabla \delta \alpha \Big) $$
$$+ \Big(\frac{\rho_d}{\rho_c} - 1 \Big) \nabla \cdot \Big(\frac{\mu_m^t}{\rho_c} \delta_\alpha \nabla \alpha \Big) + \Big(\frac{\rho_d}{\rho_c} - 1 \Big) \nabla \cdot \Big(\frac{\mu_m^t}{\rho_c} \alpha \nabla \delta \alpha \Big) \Big) = 0. \quad (11)$$

Decomposing the objective function into contributions from the boundary, Γ, and interior, Ω, of the domain,

$$J = \int_\Gamma J_\Gamma\, d\Gamma + \int_\Omega J_\Omega\, d\Omega, \quad (12)$$

Equation (11) can be reformulated as,

$$\int_\Gamma d\Gamma\, \Big(\mathbf{n}(\mathbf{u} \cdot \rho_m \mathbf{v}_m) + \mathbf{u}(\rho_m \mathbf{v}_m \cdot \mathbf{n}) + 2\mu_m \mathbf{n} \cdot D(\mathbf{u}) - q\rho_m \mathbf{n} + \alpha \beta \mathbf{n} + \frac{\partial J_\Gamma}{\partial \mathbf{v}_m} \Big) \cdot \delta \mathbf{v}_m$$
$$- \int_\Gamma d\Gamma \big(2\mu_m \mathbf{n} \cdot D(\delta \mathbf{v}_m) \cdot \mathbf{u} + 2\delta_{v_m} \mu_d \mathbf{n} \cdot D(\mathbf{v}_m) \cdot \mathbf{u} - 2\delta_{v_m} \mu_d \mathbf{n} \cdot D(\mathbf{u}) \cdot \mathbf{v}_m \big)$$
$$+ \int_\Omega d\Omega \Big(- \nabla \mathbf{u} \cdot (\rho_m \mathbf{v}_m) - (\rho_m \mathbf{v}_m \cdot \nabla) \mathbf{u} - \nabla \cdot (2\mu_m D(\mathbf{u}))$$
$$+ \aleph \rho_m \mathbf{u} + \rho_m \nabla q - \alpha \nabla \beta + \frac{\partial J_\Omega}{\partial \mathbf{v}_m} \Big) \cdot \delta \mathbf{v}_m - \int_\Omega d\Omega\, \nabla \cdot (2\delta_{v_m} \mu_d D(\mathbf{u})) \cdot \mathbf{v}_m$$
$$+ \int_\Gamma d\Gamma \Big(\rho_m \mathbf{u} \cdot \mathbf{n} + \frac{\partial J_\Gamma}{\partial p_m} \Big) \delta p_m + \int_\Omega d\Omega \Big(- \nabla \cdot \rho_m \mathbf{u} + \frac{\partial J_\Omega}{\partial p_m} \Big) \delta p_m$$
$$+ \int_\Gamma d\Gamma \Big((\rho_d - \rho_c)\big(\mathbf{u}(\mathbf{v}_m \cdot \mathbf{n}) \cdot \mathbf{v}_m + p_m \mathbf{u} \cdot \mathbf{n} - q \mathbf{v}_m \cdot \mathbf{n} \big) + \beta \mathbf{v}_m \cdot \mathbf{n} + \frac{\mu_m^t}{\rho_c}(\mathbf{n} \cdot \nabla)\beta$$
$$+ \frac{\mu_m^t}{\rho_c}\Big(\frac{\rho_d}{\rho_c} - 1 \Big)\big(\beta(\mathbf{n} \cdot \nabla)\alpha - \alpha(\mathbf{n} \cdot \nabla)\beta \big) + \frac{\partial J_\Gamma}{\partial \alpha} \Big) \delta \alpha$$
$$+ \int_\Gamma d\Gamma\, \big(\mathbf{u} \cdot \mathbf{n} \delta_\alpha(\alpha \rho_d \mathbf{v}_{dj} \mathbf{v}_{dj}) + \beta\, \delta_\alpha(\alpha \mathbf{v}_{dj}) \cdot \mathbf{n} - 2\delta_\alpha \mu_d \mathbf{n} \cdot D(\mathbf{v}_m) \cdot \mathbf{u} + 2\delta_\alpha \mu_d \mathbf{n} \cdot D(\mathbf{u}) \cdot \mathbf{v}_m \big)$$
$$- \int_\Gamma d\Gamma\, \beta \frac{\mu_m^t}{\rho_c}\Big(1 - \alpha\Big(\frac{\rho_d}{\rho_c} - 1 \Big) \Big)(\mathbf{n} \cdot \nabla)\delta \alpha$$
$$+ \int_\Omega d\Omega \Big((\rho_d - \rho_c)\big(-(\mathbf{v}_m \cdot \nabla)\mathbf{u} \cdot \mathbf{v}_m - p_m \nabla \cdot \mathbf{u} + \mathbf{u} \cdot (\aleph \mathbf{v}_m - \mathbf{g}) + (\mathbf{v}_m \cdot \nabla)q \big)$$
$$- (\mathbf{v}_m \cdot \nabla)\beta + \frac{\mu_m^t}{\rho_c}\Big(\Big(\frac{\rho_d}{\rho_c} - 1 \Big)\big(\nabla \cdot (\alpha \nabla \beta) - \nabla \alpha \cdot \nabla \beta \big) - \nabla \cdot \nabla \beta \Big) + \frac{\partial J_\Omega}{\partial \alpha} \Big) \delta \alpha$$
$$- \int_\Omega d\Omega\, \Big(\nabla \cdot \mathbf{u}\, \delta_\alpha(\alpha \rho_d \mathbf{v}_{dj} \mathbf{v}_{dj}) + \delta_\alpha(\alpha \mathbf{v}_{dj}) \cdot \nabla \beta + \nabla \cdot (2\delta_\alpha \mu_d D(\mathbf{u})) \cdot \mathbf{v}_m \Big) = 0. \quad (13)$$

Derivation of Equation (13) can be found in Appendix D. In order to satisfy Equation (13) in general, the integrals must vanish individually. The adjoint drift flux equations are deduced from the integrals over the interior of the domain:

$$-\nabla \mathbf{u} \cdot (\rho_m \mathbf{v}_m) - (\rho_m \mathbf{v}_m \cdot \nabla)\mathbf{u} - \nabla \cdot (2\mu_m D(\mathbf{u})) = -\rho_m \nabla q + \alpha \nabla \beta - \aleph \rho_m \mathbf{u} - \frac{\partial J_\Omega}{\partial \mathbf{v}_m}$$
$$+ \nabla \cdot \left(2\frac{\partial \mu_d}{\partial \mathbf{v}_m} D(\mathbf{u})\right) \cdot \mathbf{v}_m, \quad (14a)$$

$$\nabla \cdot (\rho_m \mathbf{u}) = \frac{\partial J_\Omega}{\partial p_m}, \quad (14b)$$

$$-\left(\mathbf{v}_m + \frac{\partial}{\partial \alpha}(\alpha \mathbf{v}_{dj}) + \frac{\mu_m^t}{\rho_c}\left(\frac{\rho_d}{\rho_c} - 1\right)\nabla \alpha\right) \cdot \nabla \beta = \nabla \cdot \frac{\mu_m^t}{\rho_c}\left(1 - \alpha\left(\frac{\rho_d}{\rho_c} - 1\right)\right)\nabla \beta$$
$$+ S_1 + S_2 - \frac{\partial J_\Omega}{\partial \alpha}, \quad (14c)$$

where:

$$S_1 = (\rho_d - \rho_c)((\mathbf{v}_m \cdot \nabla)(\mathbf{u} \cdot \mathbf{v}_m - q) - \mathbf{u} \cdot (\aleph \mathbf{v}_m - \mathbf{g})) + \nabla \cdot \left(2\frac{\partial \mu_d}{\partial \alpha} D(\mathbf{u})\right) \cdot \mathbf{v}_m, \quad (15a)$$

$$S_2 = \left((\rho_d - \rho_c) p_m + \frac{\partial}{\partial \alpha}(\alpha \rho_d \mathbf{v}_{dj} \mathbf{v}_{dj})\right)\nabla \cdot \mathbf{u}, \quad (15b)$$

and the boundary conditions for the adjoint variables are deduced from the surface integrals:

$$\int_\Gamma d\Gamma \left(\mathbf{n}(\mathbf{u} \cdot \rho_m \mathbf{v}_m) + \rho_m \mathbf{v}_m \cdot \mathbf{n}\mathbf{u} + 2\mu_m \mathbf{n} \cdot D(\mathbf{u}) - q\rho_m \mathbf{n} + \alpha \beta \mathbf{n} + \frac{\partial J_\Gamma}{\partial \mathbf{v}_m}\right.$$
$$\left. + 2\frac{\partial \mu_d}{\partial \mathbf{v}_m} \mathbf{n} \cdot (D(\mathbf{v}_m) \cdot \mathbf{u} - D(\mathbf{u}) \cdot \mathbf{v}_m)\right) \cdot \delta \mathbf{v}_m - \int_\Gamma d\Gamma 2\mu_m \mathbf{n} \cdot D(\delta \mathbf{v}_m) \cdot \mathbf{u} = 0, \quad (16a)$$

$$\int_\Gamma d\Gamma \left(\rho_m u_n + \frac{\partial J_\Gamma}{\partial p_m}\right)\delta p_m = 0, \quad (16b)$$

$$\int_\Gamma d\Gamma \left(\left(\mathbf{v}_m \cdot \mathbf{n} + \frac{\partial}{\partial \alpha}(\alpha \mathbf{v}_{dj}) \cdot \mathbf{n} + \frac{\mu_m^t}{\rho_c}\left(\frac{\rho_d}{\rho_c} - 1\right)(\mathbf{n} \cdot \nabla)\alpha\right)\beta\right.$$
$$\left. + \frac{\mu_m^t}{\rho_c}\left(1 - \alpha\left(\frac{\rho_d}{\rho_c} - 1\right)\right)(\mathbf{n} \cdot \nabla)\beta + C_1 + C_2 + \frac{\partial J_\Gamma}{\partial \alpha}\right)\delta \alpha \quad (16c)$$
$$- \int_\Gamma d\Gamma \beta \frac{\mu_m^t}{\rho_c}\left(1 - \alpha\left(\frac{\rho_d}{\rho_c} - 1\right)\right)(\mathbf{n} \cdot \nabla)\delta \alpha = 0,$$

where:

$$C_1 = \left((\rho_d - \rho_c)(\mathbf{u} \cdot \mathbf{v}_m - q)\mathbf{v}_m + 2\frac{\partial \mu_d}{\partial \alpha}(D(\mathbf{u}) \cdot \mathbf{v}_m)\right) \cdot \mathbf{n}, \quad (17a)$$

$$C_2 = \left(\left((\rho_d - \rho_c) p_m + \frac{\partial}{\partial \alpha}(\alpha \rho_d \mathbf{v}_{dj} \mathbf{v}_{dj})\right)\mathbf{u} - 2\frac{\partial \mu_d}{\partial \alpha}(D(\mathbf{v}_m) \cdot \mathbf{u})\right) \cdot \mathbf{n} \quad (17b)$$

and $u_n = \mathbf{u} \cdot \mathbf{n}$ is the normal component of the adjoint velocity. This is the general form of the adjoint equation system for the steady state drift flux equations with Darcy porosity term and frozen turbulence.

3. Application to Wall Bounded Flows

Thus far in the paper we have presented the optimisation problem in as generic a way as possible. To proceed further with the derivation we now need to derive expressions for the boundary conditions, objective function and slip velocity. We will examine these for the case of wall-bounded or ducted flows, for which there is no contribution to the objective function from the interior of the domain. So, in the cases where the objective function only involves integrals over the surface of the flow domain rather than over its interior, the adjoint equations reduce to:

$$-\nabla \mathbf{u} \cdot (\rho_m \mathbf{v}_m) - (\rho_m \mathbf{v}_m \cdot \nabla)\mathbf{u} - \nabla \cdot (2\mu_m D(\mathbf{u})) = -\rho_m \nabla q + \alpha \nabla \beta - \aleph \rho_m \mathbf{u}$$
$$+ \nabla \cdot \left(2\frac{\partial \mu_d}{\partial \mathbf{v}_m} D(\mathbf{u})\right) \cdot \mathbf{v}_m, \quad (18a)$$

$$\nabla \cdot (\rho_m \mathbf{u}) = 0, \quad (18b)$$

$$-\left(\mathbf{v}_m + \frac{\partial}{\partial \alpha}(\alpha \mathbf{v}_{dj}) + \frac{\mu_m^t}{\rho_c}\left(\frac{\rho_d}{\rho_c} - 1\right)\nabla \alpha\right) \cdot \nabla \beta = \nabla \cdot \frac{\mu_m^t}{\rho_c}\left(1 - \alpha\left(\frac{\rho_d}{\rho_c} - 1\right)\right)\nabla \beta$$
$$+ S_1. \quad (18c)$$

These equations no longer depend on the objective function, so when switching from one optimisation objective to another, they remain unchanged and only the boundary conditions have to be adapted to the specific objective function. Note that as a result of Equation (18b), $\nabla \cdot \mathbf{u} = 0$ [22] and, therefore, $S_2 = 0$.

For the adjoint boundary conditions, the terms in Equation (16a) involving μ_m can be rewritten as,

$$\int_\Gamma d\Gamma \, 2\mu_m \mathbf{n} \cdot (D(\mathbf{u}) \cdot \delta \mathbf{v}_m - D(\delta \mathbf{v}_m) \cdot \mathbf{u}) = \int_\Gamma d\Gamma \, \mu_m ((\mathbf{n} \cdot \nabla)\mathbf{u} \cdot \delta \mathbf{v}_m - (\mathbf{n} \cdot \nabla)\delta \mathbf{v}_m \cdot \mathbf{u}) \quad (19)$$

(Ref. [19]) and therefore the adjoint boundary conditions, Equation (16), reduce to:

$$\int_\Gamma d\Gamma \left(\mathbf{n}(\mathbf{u} \cdot \rho_m \mathbf{v}_m) + \rho_m \mathbf{v}_m \cdot \mathbf{n}\mathbf{u} + \mu_m (\mathbf{n} \cdot \nabla)\mathbf{u} - q\rho_m \mathbf{n} + \alpha \beta \mathbf{n} + \frac{\partial J_\Gamma}{\partial \mathbf{v}_m}\right.$$
$$\left. + 2\frac{\partial \mu_d}{\partial \mathbf{v}_m} \mathbf{n} \cdot (D(\mathbf{v}_m) \cdot \mathbf{u} - D(\mathbf{u}) \cdot \mathbf{v}_m)\right) \cdot \delta \mathbf{v}_m - \int_\Gamma d\Gamma \, \mu_m (\mathbf{n} \cdot \nabla)\delta \mathbf{v}_m \cdot \mathbf{u} = 0, \quad (20a)$$

$$\int_\Gamma d\Gamma \left(\rho_m u_n + \frac{\partial J_\Gamma}{\partial p_m}\right) \delta p_m = 0, \quad (20b)$$

$$\int_\Gamma d\Gamma \left(\left(\mathbf{v}_m \cdot \mathbf{n} + \frac{\partial}{\partial \alpha}(\alpha \mathbf{v}_{dj}) \cdot \mathbf{n} + \frac{\mu_m^t}{\rho_c}\left(\frac{\rho_d}{\rho_c} - 1\right)(\mathbf{n} \cdot \nabla)\alpha\right)\beta\right.$$
$$\left. + \frac{\mu_m^t}{\rho_c}\left(1 - \alpha\left(\frac{\rho_d}{\rho_c} - 1\right)\right)(\mathbf{n} \cdot \nabla)\beta + C_1 + C_2 + \frac{\partial J_\Gamma}{\partial \alpha}\right)\delta \alpha$$
$$- \int_\Gamma d\Gamma \beta \frac{\mu_m^t}{\rho_c}\left(1 - \alpha\left(\frac{\rho_d}{\rho_c} - 1\right)\right)(\mathbf{n} \cdot \nabla)\delta \alpha = 0. \quad (20c)$$

In order to determine the boundary conditions of the adjoint variables, the boundary conditions imposed on the primal variables are listed in Table 1. We will derive expressions for the three main boundary conditions.

Table 1. Primal boundary conditions.

	u_m	α	p_m
Inlet	fixed value	fixed value	zero gradient
Wall	zero	zero gradient	zero gradient
Outlet	zero gradient	zero gradient	zero

3.1. Adjoint Boundary Conditions at the Inlet

At an inlet, the primal velocity and dispersed-phase volume fraction are usually fixed, so,

$$\delta \mathbf{v}_m = 0 \text{ and } \delta \alpha = 0. \tag{21}$$

The first integrals in Equations (20a) and (20c) therefore go to zero and Equation (20) reduces to:

$$\int_\Gamma d\Gamma \mu_m (\mathbf{n} \cdot \nabla) \delta \mathbf{v}_m \cdot \mathbf{u} = 0, \tag{22a}$$

$$\int_\Gamma d\Gamma \left(\rho_m u_n + \frac{\partial J_\Gamma}{\partial p_m} \right) \delta p_m = 0, \tag{22b}$$

$$\int_\Gamma d\Gamma \beta \frac{\mu_m^t}{\rho_c} \left(1 - \alpha \left(\frac{\rho_d}{\rho_c} - 1 \right) \right) (\mathbf{n} \cdot \nabla) \delta \alpha = 0. \tag{22c}$$

When both fluids are incompressible, $\nabla \cdot \mathbf{v}_m = 0$ [22], and as $\delta \mathbf{v}_{mt} = 0$ along the inlet, $(\mathbf{n} \cdot \nabla) \delta \mathbf{v}_m = (\mathbf{n} \cdot \nabla) \delta \mathbf{v}_{mt}$ [19], where $\mathbf{v}_{mt}$ is the tangential component of the mixture velocity. Hence, Equation (22) reduces to:

$$\int_\Gamma d\Gamma \mu_m (\mathbf{n} \cdot \nabla) \delta \mathbf{v}_{mt} \cdot \mathbf{u}_t = 0, \tag{23a}$$

$$\int_\Gamma d\Gamma \left(\rho_m u_n + \frac{\partial J_\Gamma}{\partial p_m} \right) \delta p_m = 0, \tag{23b}$$

$$\int_\Gamma d\Gamma \beta \frac{\mu_m^t}{\rho_c} \left(1 - \alpha \left(\frac{\rho_d}{\rho_c} - 1 \right) \right) (\mathbf{n} \cdot \nabla) \delta \alpha = 0, \tag{23c}$$

where $\mathbf{u}_t$ is the tangential component of the adjoint velocity, from which we deduce the boundary conditions for the adjoint variables at the inlet to be:

$$\mathbf{u}_t = 0, \tag{24a}$$

$$u_n = -\frac{1}{\rho_m} \frac{\partial J_\Gamma}{\partial p_m}, \tag{24b}$$

$$\beta = 0 \iff \mu_m^t \neq 0. \tag{24c}$$

Note that these derivations do not impose a condition for q. Since q enters the adjoint drift flux equations in a manner similar to the way p_m enters the primal drift flux equations, the zero gradient boundary condition of p_m at the inlet is applied to q as well,

$$(\mathbf{n} \cdot \nabla) q = 0. \tag{25}$$

3.2. Adjoint Boundary Conditions at the Wall

At a wall, typical primal conditions are zero velocity and zero gradient of the dispersed-phase volume fraction. Therefore, we have,

$$\mathbf{v}_m = 0, \; \delta \mathbf{v}_m = 0 \text{ and } (\mathbf{n} \cdot \nabla) \delta \alpha = 0. \tag{26}$$

The first integral in Equation (20a) and the second integral in Equation (20c) therefore go to zero and the terms in the first integral in Equation (20c), containing $\mathbf{v}_m$, go to zero. Equation (20) therefore reduces to:

$$\int_\Gamma d\Gamma \mu_m (\mathbf{n} \cdot \nabla) \delta \mathbf{v}_m \cdot \mathbf{u} = 0, \tag{27a}$$

$$\int_\Gamma d\Gamma \left(\rho_m u_n + \frac{\partial J_\Gamma}{\partial p_m} \right) \delta p_m = 0, \tag{27b}$$

$$\int_\Gamma d\Gamma \left(\left(\frac{\partial}{\partial \alpha}(\alpha \mathbf{v}_{dj}) \cdot \mathbf{n} + \frac{\mu_m^t}{\rho_c} \left(\frac{\rho_d}{\rho_c} - 1 \right) (\mathbf{n} \cdot \nabla) \alpha \right) \beta \right.$$
$$\left. + \frac{\mu_m^t}{\rho_c} \left(1 - \alpha \left(\frac{\rho_d}{\rho_c} - 1 \right) \right) (\mathbf{n} \cdot \nabla) \beta + C_2 + \frac{\partial J_\Gamma}{\partial \alpha} \right) \delta \alpha = 0. \tag{27c}$$

As at the inlet, the primal velocity does not diverge and $\delta \mathbf{v}_{mt} = 0$ along the wall, so Equation (27) reduces to:

$$\int_\Gamma d\Gamma \mu_m (\mathbf{n} \cdot \nabla) \delta \mathbf{v}_{mt} \cdot \mathbf{u}_t = 0, \tag{28a}$$

$$\int_\Gamma d\Gamma \left(\rho_m u_n + \frac{\partial J_\Gamma}{\partial p_m} \right) \delta p_m = 0, \tag{28b}$$

$$\int_\Gamma d\Gamma \left(\left(\frac{\partial}{\partial \alpha}(\alpha \mathbf{v}_{dj}) \cdot \mathbf{n} + \frac{\mu_m^t}{\rho_c} \left(\frac{\rho_d}{\rho_c} - 1 \right) (\mathbf{n} \cdot \nabla) \alpha \right) \beta \right.$$
$$\left. + \frac{\mu_m^t}{\rho_c} \left(1 - \alpha \left(\frac{\rho_d}{\rho_c} - 1 \right) \right) (\mathbf{n} \cdot \nabla) \beta + C_2 + \frac{\partial J_\Gamma}{\partial \alpha} \right) \delta \alpha = 0, \tag{28c}$$

from which we deduce the boundary conditions for the adjoint variables at the wall to be:

$$\mathbf{u}_t = 0, \tag{29a}$$

$$u_n = -\frac{1}{\rho_m} \frac{\partial J_\Gamma}{\partial p_m}, \tag{29b}$$

$$\left(\frac{\partial}{\partial \alpha}(\alpha \mathbf{v}_{dj}) \cdot \mathbf{n} + \frac{\mu_m^t}{\rho_c} \left(\frac{\rho_d}{\rho_c} - 1 \right) (\mathbf{n} \cdot \nabla) \alpha \right) \beta$$
$$+ \frac{\mu_m^t}{\rho_c} \left(1 - \alpha \left(\frac{\rho_d}{\rho_c} - 1 \right) \right) (\mathbf{n} \cdot \nabla) \beta = -C_2 - \frac{\partial J_\Gamma}{\partial \alpha}. \tag{29c}$$

Equation (29c) is used to determine β and, as at the inlet, Equation (25) applies.

3.3. Adjoint Boundary Conditions at the Outlet

At an outlet, typical primal conditions are zero pressure and zero gradient of velocity and dispersed-phase volume fraction. Therefore, we have,

$$\delta p_m = 0, \ (\mathbf{n} \cdot \nabla) \delta \mathbf{v}_m = 0 \text{ and } (\mathbf{n} \cdot \nabla) \delta \alpha = 0. \tag{30}$$

The second integral in Equations (20a) and (20c) therefore goes to zero and, with $\delta p_m = 0$, Equation (20b) is identically fulfilled. The remaining terms in Equation (20) are the first integrals in Equations (20a) and (20c), which can be made to go to zero by enforcing the integrands to vanish:

$$\mathbf{n}(\mathbf{u} \cdot \rho_m \mathbf{v}_m) + \rho_m \mathbf{v}_m \cdot \mathbf{n}\mathbf{u} + \mu_m (\mathbf{n} \cdot \nabla)\mathbf{u} - q\rho_m \mathbf{n} + \alpha\beta\mathbf{n} + \frac{\partial J_\Gamma}{\partial \mathbf{v}_m}$$
$$-2\frac{\partial \mu_d}{\partial \mathbf{v}_m}\mathbf{n} \cdot D(\mathbf{u}) \cdot \mathbf{v}_m = 0, \quad (31\text{a})$$

$$\left(\mathbf{v}_m \cdot \mathbf{n} + \frac{\partial}{\partial \alpha}(\alpha \mathbf{v}_{dj}) \cdot \mathbf{n} + \frac{\mu_m^t}{\rho_c}\left(\frac{\rho_d}{\rho_c} - 1\right)(\mathbf{n} \cdot \nabla)\alpha\right)\beta$$
$$+\frac{\mu_m^t}{\rho_c}\left(1 - \alpha\left(\frac{\rho_d}{\rho_c} - 1\right)\right)(\mathbf{n} \cdot \nabla)\beta + C_1 + C_2 + \frac{\partial J_\Gamma}{\partial \alpha} = 0. \quad (31\text{b})$$

Note that the term containing $D(\mathbf{v}_m) = 0$, because $(\mathbf{n} \cdot \nabla)\mathbf{v}_m = 0$. Decomposing Equation (31a) into its normal and tangential components yields:

$$\rho_m \mathbf{u} \cdot \mathbf{v}_m + \rho_m u_n \mathbf{v}_m \cdot \mathbf{n} + \mu_m(\mathbf{n} \cdot \nabla)u_n - \rho_m q + \alpha\beta + \frac{\partial J_\Gamma}{\partial \mathbf{v}_m \cdot \mathbf{n}}$$
$$-2\frac{\partial \mu_d}{\partial \mathbf{v}_m}\mathbf{n} \cdot D(\mathbf{u}) \cdot \mathbf{v}_m \cdot \mathbf{n} = 0, \quad (32\text{a})$$

$$\rho_m \mathbf{v}_m \cdot \mathbf{n}\mathbf{u}_t + \mu_m(\mathbf{n} \cdot \nabla)\mathbf{u}_t + \frac{\partial J_\Gamma}{\partial \mathbf{v}_{mt}} - 2\frac{\partial \mu_d}{\partial \mathbf{v}_m}\mathbf{n} \cdot D(\mathbf{u}) \cdot \mathbf{v}_{mt} = 0. \quad (32\text{b})$$

Equations (31b), (32a) and (32b) are used to determine β, q and u_t, respectively. Since u_n is prescribed at the inlet, the adjoint continuity equation, Equation (18b), is used to calculate u_n at the outlet, Φ_Γ. The boundary conditions for the adjoint variables at the outlet are summarised as:

$$\rho_m \mathbf{v}_m \cdot \mathbf{n}\mathbf{u}_t + \mu_m(\mathbf{n} \cdot \nabla)\mathbf{u}_t - 2\frac{\partial \mu_d}{\partial \mathbf{v}_m}\mathbf{n} \cdot D(\mathbf{u}) \cdot \mathbf{v}_{mt} = -\frac{\partial J_\Gamma}{\partial \mathbf{v}_{mt}}, \quad (33\text{a})$$

$$u_n = \Phi_\Gamma, \quad (33\text{b})$$

$$\left(\mathbf{v}_m \cdot \mathbf{n} + \frac{\partial}{\partial \alpha}(\alpha \mathbf{v}_{dj}) \cdot \mathbf{n} + \frac{\mu_m^t}{\rho_c}\left(\frac{\rho_d}{\rho_c} - 1\right)(\mathbf{n} \cdot \nabla)\alpha\right)\beta$$
$$+\frac{\mu_m^t}{\rho_c}\left(1 - \alpha\left(\frac{\rho_d}{\rho_c} - 1\right)\right)(\mathbf{n} \cdot \nabla)\beta = -C_1 - C_2 - \frac{\partial J_\Gamma}{\partial \alpha}, \quad (33\text{c})$$

$$\mathbf{u} \cdot \mathbf{v}_m + u_n \mathbf{v}_m \cdot \mathbf{n} + \nu_m(\mathbf{n} \cdot \nabla)u_n + \frac{\alpha\beta}{\rho_m} + \frac{1}{\rho_m}\frac{\partial J_\Gamma}{\partial \mathbf{v}_m \cdot \mathbf{n}}$$
$$-\frac{2}{\rho_m}\frac{\partial \mu_d}{\partial \mathbf{v}_m}\mathbf{n} \cdot D(\mathbf{u}) \cdot \mathbf{v}_m \cdot \mathbf{n} = q, \quad (33\text{d})$$

where $\nu_m = \frac{\mu_m}{\rho_m}$ is the mixture kinematic viscosity. A summary of the boundary conditions for the adjoint variables is presented in Table 2.

Table 2. Adjoint boundary conditions for ducted flows.

	u_t	u_n	β	q
Inlet	zero	Equation (24b)	zero	zero gradient
Wall	zero	Equation (29b)	Equation (29c)	zero gradient
Outlet	Equation (33a)	Equation (33b)	Equation (33c)	Equation (33d)

4. Objective Function

The objective function is related to the dispersed-phase mass-flow rate at the boundaries of the domain,

$$J_\Gamma = \alpha \rho_d \mathbf{v}_d \cdot \mathbf{n}, \tag{34}$$

where $\alpha \rho_d$ is the dispersed-phase mass fraction and $\mathbf{v}_d \cdot \mathbf{n}$ is the dispersed-phase velocity normal to the boundary. Since the phase fraction at the inlet is specified, the objective function is defined as the mass-flow rate of solid at the outlet, and Equation (12) becomes,

$$J = \int_{\Gamma_o} J_{\Gamma_o} d\Gamma_o, \tag{35}$$

where $_o$ refers to the outlet. The derivatives of the objective function, Equation (34), with respect to the primal variables are:

$$\frac{\partial J_\Gamma}{\partial \mathbf{v}_m \cdot \mathbf{n}} = \alpha \rho_d, \tag{36a}$$

$$\frac{\partial J_\Gamma}{\partial \mathbf{v}_{mt}} = 0, \tag{36b}$$

$$\frac{\partial J_\Gamma}{\partial \alpha} = \rho_d \left(\mathbf{v}_m + \mathbf{v}_{dj} + \alpha \frac{\partial \mathbf{v}_{dj}}{\partial \alpha} \right) \cdot \mathbf{n}, \tag{36c}$$

$$\frac{\partial J_\Gamma}{\partial p_m} = 0. \tag{36d}$$

Derivation of Equation (36c) can be found in Appendix E. Using these derivatives, the adjoint boundary conditions at an inlet reduces to:

$$u_t = 0, \tag{37a}$$
$$u_n = 0, \tag{37b}$$
$$\beta = 0 \iff \mu_m^t \neq 0, \tag{37c}$$
$$(\mathbf{n} \cdot \nabla)q = 0. \tag{37d}$$

At a wall, there is no contribution from the objective function, so:

$$u_t = 0, \tag{38a}$$
$$u_n = 0, \tag{38b}$$

$$\left(\frac{\partial}{\partial \alpha}(\alpha \mathbf{v}_{dj}) \right) \cdot \mathbf{n} + \frac{\mu_m^t}{\rho_c} \left(\frac{\rho_d}{\rho_c} - 1 \right) (\mathbf{n} \cdot \nabla) \alpha \right) \beta$$
$$+ \frac{\mu_m^t}{\rho_c} \left(1 - \alpha \left(\frac{\rho_d}{\rho_c} - 1 \right) \right) (\mathbf{n} \cdot \nabla) \beta = 0, \tag{38c}$$

$$(\mathbf{n} \cdot \nabla)q = 0. \tag{38d}$$

Note that $C_2 = 0$ because $\mathbf{u} = 0$. At an outlet, to satisfy the adjoint continuity equation, Equation (18b), $u_n = 0$, so:

$$\mathbf{v}_m \cdot \mathbf{n} u_t + \nu_m (\mathbf{n} \cdot \nabla) \mathbf{u}_t - \frac{2}{\rho_m} \frac{\partial \mu_d}{\partial \mathbf{v}_m} \mathbf{n} \cdot D(\mathbf{u}) \cdot \mathbf{v}_{mt} = 0, \tag{39a}$$

$$u_n = 0, \tag{39b}$$

$$\left(\mathbf{v}_m \cdot \mathbf{n} + \frac{\partial}{\partial \alpha}(\alpha \mathbf{v}_{dj}) \cdot \mathbf{n} + \frac{\mu_m^t}{\rho_c} \left(\frac{\rho_d}{\rho_c} - 1 \right) (\mathbf{n} \cdot \nabla) \alpha \right) \beta$$
$$+ \frac{\mu_m^t}{\rho_c} \left(1 - \alpha \left(\frac{\rho_d}{\rho_c} - 1 \right) \right) (\mathbf{n} \cdot \nabla) \beta = -C_1 - \frac{\partial J_\Gamma}{\partial \alpha}, \tag{39c}$$

$$\mathbf{u} \cdot \mathbf{v}_m + \nu_m (\mathbf{n} \cdot \nabla) u_n + \frac{\alpha \beta}{\rho_m} + \frac{1}{\rho_m} \frac{\partial J_\Gamma}{\partial \mathbf{v}_m \cdot \mathbf{n}} - \frac{2}{\rho_m} \frac{\partial \mu_d}{\partial \mathbf{v}_m} \mathbf{n} \cdot D(\mathbf{u}) \cdot \mathbf{v}_m \cdot \mathbf{n} = q. \tag{39d}$$

Note that $C_2 = 0$ because $u_n = 0$ and $D(\mathbf{v}_m) = 0$ when $(\mathbf{n} \cdot \nabla) \mathbf{v}_m = 0$. A summary of the adjoint boundary conditions, using the objective function defined in Equation (34), is presented in Table 3.

Table 3. Adjoint boundary conditions, using objective function Equation (34).

	u_t	u_n	β	q
Inlet	zero	zero	zero	zero gradient
Wall	zero	zero	Equation (38c)	zero gradient
Outlet	Equation (39a)	zero	Equation (39c)	Equation (39d)

5. Settling Velocity

The equations thus far have been derived for the most general case in which the settling (drift) velocity has not been specified. Of course the settling velocity is key to the behaviour of the drift flux model, and incorporates much of the physics of the multiphase system. Here we will derive the appropriate additional equations for two common settling velocity models, vis. the Dahl [23] and Takacs [24] models.

5.1. Dahl Model

In this formulation $\mathbf{v}_{dj}$ is modelled using,

$$\mathbf{v}_{dj} = \mathbf{v}_0 10^{-k\alpha}, \tag{40}$$

where $\mathbf{v}_0$ is the maximum theoretical settling velocity and k is a settling parameter, and its partial derivative with respect to α is given by,

$$\frac{\partial \mathbf{v}_{dj}}{\partial \alpha} = -k \ln 10\, \mathbf{v}_{dj}. \tag{41}$$

5.2. Takacs Model

In this formulation $\mathbf{v}_{dj}$ is modelled using,

$$\mathbf{v}_{dj} = \mathbf{v}_0 \left(e^{-a(\alpha - \alpha_r)} - e^{-a_1(\alpha - \alpha_r)} \right), \tag{42a}$$

$$0 \leq \mathbf{v}_{dj} \leq \mathbf{v}_{00}, \tag{42b}$$

where:

- a is the hindered settling parameter,
- a_1 is the flocculent settling parameter,

- α_r is the volume fraction of non-settleable solids at the inlet and
- v_{00} is the maximum practical settling velocity,

and its partial derivative with respect to α is given by,

$$\frac{\partial v_{dj}}{\partial \alpha} = v_0 \left(-ae^{-a(\alpha-\alpha_r)} + a_1 e^{-a_1(\alpha-\alpha_r)} \right). \tag{43}$$

For both models the partial derivative of αv_{dj} with respect to α is given by,

$$\frac{\partial}{\partial \alpha}(\alpha v_{dj}) = v_{dj} + \alpha \frac{\partial v_{dj}}{\partial \alpha}. \tag{44}$$

6. Conclusions

In this paper we have derived, for the first time, the adjoint equations based on the drift flux model for dispersed multiphase flow. In addition to the adjoint drift flux equations themselves we have presented the adjoint boundary conditions for the common boundary conditions (Inlet, Outlet, Wall), as well as a treatment of the generic objective function, and specific formulations corresponding to the common settling velocity models proposed by Dahl [23] and Takacs [24]. From these elements a full adjoint set of equations can be derived for any specific ducted flow problem, and of course implemented in an appropriate numerical code. This of course also presents many, largely numerical and coding, challenges, and this will be the subject of subsequent papers.

Author Contributions: Formal analysis, S.G.; Funding acquisition, D.S.J.; Investigation, S.G. and G.R.T.; Methodology, D.S.J.; Project administration, G.R.T.; Writing–original draft, S.G.; Writing–review and editing, D.S.J. and G.R.T. All authors have read and agreed to the published version of the manuscript.

Funding: This research was funded by the STREAM CDT under grant number EP/L015412/1 and InnovateUK under KTP grant KTP010621.

Conflicts of Interest: There is no conflict of interest for Hydro International Ltd. in the publication of the paper. Hydro International holds interests in the implementation and application of the methods discussed in the paper for specific scenarios and geometries, but this work will remain private and unpublished.

Appendix A. Derivation of Equation (10a)

The variation of $(R_1, R_2, R_3)^T$ with respect to v_m is calculated as,

$$\begin{aligned}\delta_{v_m}(R_1, R_2, R_3)^T &= \delta_{v_m}\Big((v_m \cdot \nabla)(\rho_m v_m) + \nabla(\rho_m p_m) - \nabla \cdot (2\mu_m D(v_m)) \\ &\quad + \nabla \cdot \left(\frac{\alpha}{1-\alpha} \frac{\rho_c \rho_d}{\rho_m} v_{dj} v_{dj} \right) - \rho_m g + \aleph \rho_m v_m \Big) \\ &= (\delta v_m \cdot \nabla)(\rho_m v_m) + (v_m \cdot \nabla)(\rho_m \delta v_m) - \nabla \cdot (2\mu_m D(\delta v_m)) \\ &\quad - \nabla \cdot (2 \delta_{v_m} \mu_m D(v_m)) + \aleph \rho_m \delta v_m.r \end{aligned} \tag{A1}$$

As stated above, μ_m is defined as the sum of the continuum, dispersed-phase and mixture turbulent viscosities,

$$\mu_m = \mu_c + \mu_d + \mu_m^t, \tag{A2}$$

where μ_c is constant, μ_d is a function of v_m and α, and μ_m^t is obtained from turbulence modelling. Equation (A1) can now be rewritten as,

$$\begin{aligned}\delta_{v_m}(R_1, R_2, R_3)^T &= (\delta v_m \cdot \nabla)(\rho_m v_m) + (v_m \cdot \nabla)(\rho_m \delta v_m) - \nabla \cdot (2\mu_m D(\delta v_m)) \\ &\quad - \nabla \cdot (2 \delta_{v_m} \mu_d D(v_m)) + \aleph \rho_m \delta v_m, \end{aligned} \tag{A3}$$

where $\delta_{v_m} \mu_m^t$ has been neglected.

Appendix B. Derivation of Equation (10g)

The variation of $(R_1, R_2, R_3)^T$ with respect to α is calculated as,

$$\delta_\alpha (R_1, R_2, R_3)^T = \delta_\alpha \bigg((\mathbf{v}_m \cdot \nabla)(\rho_m \mathbf{v}_m) + \nabla(\rho_m p_m) - \nabla \cdot (2\mu_m D(\mathbf{v}_m))$$

$$+ \nabla \cdot \left(\frac{\alpha}{1-\alpha} \frac{\rho_c \rho_d}{\rho_m} \mathbf{v}_{dj} \mathbf{v}_{dj} \right) - \rho_m \mathbf{g} + \aleph \rho_m \mathbf{v}_m \bigg)$$

$$= \delta_\alpha \big((\mathbf{v}_m \cdot \nabla)(\rho_m \mathbf{v}_m) \big) + \delta_\alpha \nabla(\rho_m p_m) - \delta_\alpha \nabla \cdot (2\mu_m D(\mathbf{v}_m))$$

$$+ \delta_\alpha \nabla \cdot A + \delta_\alpha (\rho_m (\aleph \mathbf{v}_m - \mathbf{g})), \tag{A4}$$

where

$$A = \frac{\alpha}{1-\alpha} \frac{\rho_c \rho_d}{\rho_m} \mathbf{v}_{dj} \mathbf{v}_{dj} \tag{A5}$$

and

$$\rho_m = \alpha \rho_d + (1-\alpha) \rho_c$$

$$= (1-\alpha) \rho_c \left(1 + \frac{\alpha}{1-\alpha} \frac{\rho_d}{\rho_c} \right). \tag{A6}$$

Substituting Equation (A6) into Equation (A5) and rewriting the parentheses as binomial expansions,

$$A = \alpha (1-\alpha)^{-1} \rho_c \rho_d \frac{(1-\alpha)^{-1}}{\rho_c} \left(1 + \frac{\alpha}{1-\alpha} \frac{\rho_d}{\rho_c} \right)^{-1} \mathbf{v}_{dj} \mathbf{v}_{dj}$$

$$= \alpha \rho_d \left(1 + \alpha \left(2 - \frac{1}{1-\alpha} \frac{\rho_d}{\rho_c} \right) + \cdots \right) \mathbf{v}_{dj} \mathbf{v}_{dj}. \tag{A7}$$

As $\alpha \ll 1$ and $\rho_d \approx 2\rho_c \implies \left| 2 - \frac{1}{1-\alpha} \frac{\rho_d}{\rho_c} \right| < 1$ and ignoring terms containing squared and higher powers of α,

$$A \approx \alpha \rho_d \mathbf{v}_{dj} \mathbf{v}_{dj}. \tag{A8}$$

Substituting Equations (A8) and (A2) into Equation (A4),

$$\delta_\alpha (R_1, R_2, R_3)^T \approx (\rho_d - \rho_c) \big((\mathbf{v}_m \cdot \nabla)(\delta_\alpha \mathbf{v}_m) + \nabla(\delta_\alpha p_m) + \delta_\alpha (\aleph \mathbf{v}_m - \mathbf{g}) \big)$$

$$- \nabla \cdot (2\delta_\alpha \mu_d D(\mathbf{v}_m)) + \nabla \cdot \delta_\alpha (\alpha \rho_d \mathbf{v}_{dj} \mathbf{v}_{dj}), \tag{A9}$$

where $\delta_\alpha \mu_m^t$ has been neglected.

Appendix C. Derivation of Equation (10i)

Similarly, the variation of R_5 with respect to α is calculated as,

$$\delta_\alpha R_5 = \delta_\alpha \left(\nabla \cdot (\alpha \mathbf{v}_m) + \nabla \cdot \left(\frac{\alpha \rho_c}{\rho_m} \mathbf{v}_{dj} \right) - \nabla \cdot (K \nabla \alpha) \right)$$

$$= \delta_\alpha \nabla \cdot (\alpha \mathbf{v}_m) + \delta_\alpha \nabla \cdot B - \delta_\alpha \nabla \cdot (K \nabla \alpha), \tag{A10}$$

where

$$B = \frac{\alpha \rho_c}{\rho_m} \mathbf{v}_{dj}. \tag{A11}$$

Substituting Equation (A6) into Equation (A11) and rewriting the parentheses as binomial expansions,

$$B = \alpha \rho_c \frac{(1-\alpha)^{-1}}{\rho_c} \left(1 + \frac{\alpha}{1-\alpha} \frac{\rho_d}{\rho_c}\right)^{-1} \mathbf{v}_{dj}$$

$$= \alpha \left(1 + \alpha \left(1 - \frac{1}{1-\alpha} \frac{\rho_d}{\rho_c}\right) + \cdots \right) \mathbf{v}_{dj}. \quad (A12)$$

As $\alpha \ll 1$ and $\rho_d \approx 2\rho_c \implies \left|1 - \frac{1}{1-\alpha}\frac{\rho_d}{\rho_c}\right| < 2$ and ignoring terms containing squared and higher powers of α,

$$B \approx \alpha \mathbf{v}_{dj}. \quad (A13)$$

As stated above, K is defined as the mixture eddy diffusivity,

$$\nu_m^t = \frac{\mu_m^t}{\rho_m}. \quad (A14)$$

From Equation (A6), $\frac{1}{\rho_m}$ can be written as,

$$\frac{1}{\rho_m} = \frac{1}{\rho_c}(1-\alpha)^{-1}\left(1 + \frac{\alpha}{1-\alpha}\frac{\rho_d}{\rho_c}\right)^{-1}$$

$$= \frac{1}{\rho_c}\left(1 + \alpha\left(1 - \frac{\rho_d}{\rho_c}\right)\right), \quad (A15)$$

ignoring terms containing squared and higher powers of α.

Substituting Equations (A14) and (A15) into the term containing K in Equation (A10),

$$\delta_\alpha \nabla \cdot (K \nabla \alpha) = \delta_\alpha \nabla \cdot \left(\frac{\mu_m^t}{\rho_m} \nabla \alpha\right)$$

$$= \delta_\alpha \nabla \cdot \left(\frac{\mu_m^t}{\rho_c}\left(1 + \alpha\left(1 - \frac{\rho_d}{\rho_c}\right)\right) \nabla \alpha\right)$$

$$= \nabla \cdot \left(\frac{\mu_m^t}{\rho_c}\left(1 + (\alpha + \delta\alpha)\left(1 - \frac{\rho_d}{\rho_c}\right)\right) \nabla(\alpha + \delta\alpha)\right) \quad (A16)$$

$$- \nabla \cdot \left(\frac{\mu_m^t}{\rho_c}\left(1 + \alpha\left(1 - \frac{\rho_d}{\rho_c}\right)\right) \nabla \alpha\right)$$

$$= \nabla \cdot \left(\frac{\mu_m^t}{\rho_c}\delta\alpha\left(1 - \frac{\rho_d}{\rho_c}\right) \nabla \alpha\right) + \nabla \cdot \left(\frac{\mu_m^t}{\rho_c}\nabla \delta\alpha\right)$$

$$+ \nabla \cdot \left(\frac{\mu_m^t}{\rho_c}\alpha\left(1 - \frac{\rho_d}{\rho_c}\right) \nabla \delta\alpha\right),$$

ignoring the term containing $\delta\alpha\nabla\delta\alpha$, because when substitued into Equation (9) becomes terms containing squared powers of $\delta\alpha$. Equation (A10) now becomes,

$$\delta_\alpha R_5 \approx \nabla \cdot (\delta\alpha \mathbf{v}_m) + \nabla \cdot \delta_\alpha(\alpha \mathbf{v}_{dj}) - \nabla \cdot \left(\frac{\mu_m^t}{\rho_c}\nabla \delta\alpha\right)$$

$$+ \frac{\mu_m^t}{\rho_c}\left(\frac{\rho_d}{\rho_c} - 1\right) \nabla \cdot (\delta\alpha \nabla \alpha) + \frac{\mu_m^t}{\rho_c}\left(\frac{\rho_d}{\rho_c} - 1\right) \nabla \cdot (\alpha \nabla \delta\alpha), \quad (A17)$$

where $\delta_\alpha \mu_m^t$ has been neglected.

Appendix D. Derivation of Equation (13)

Decomposing the objective function into contributions from the boundary and interior of the domain, according to Equation (12), the terms in Equation (11) can be written as follows. The variations of the objective function can be written as,

$$\delta_{\mathbf{v}_m} J = \int_\Gamma d\Gamma \, \frac{\partial J_\Gamma}{\partial \mathbf{v}_m} \cdot \delta \mathbf{v}_m + \int_\Omega d\Omega \, \frac{\partial J_\Omega}{\partial \mathbf{v}_m} \cdot \delta \mathbf{v}_m, \tag{A18}$$

$$\delta_{p_m} J = \int_\Gamma d\Gamma \, \frac{\partial J_\Gamma}{\partial p_m} \delta p_m + \int_\Omega d\Omega \, \frac{\partial J_\Omega}{\partial p_m} \delta p_m \tag{A19}$$

and

$$\delta_\alpha J = \int_\Gamma d\Gamma \, \frac{\partial J_\Gamma}{\partial \alpha} \delta \alpha + \int_\Omega d\Omega \, \frac{\partial J_\Omega}{\partial \alpha} \delta \alpha. \tag{A20}$$

Applying the product rule, divergence theorem and continuity equation, and using the Einstein notation for clarity, the terms containing $\mathbf{u}$, $\mathbf{v}_m$ and ∇ can be written as,

$$\begin{aligned}
\int_\Omega d\Omega \, \mathbf{u} \cdot (\delta \mathbf{v}_m \cdot \nabla)(\rho_m \mathbf{v}_m) &= \int_\Omega d\Omega \, u_k \delta v_{mi} \frac{\partial}{\partial x_i}(\rho_{mk} v_{mk}) \\
&= \int_\Omega d\Omega \, \frac{\partial}{\partial x_i}(u_k \rho_{mk} v_{mk} \delta v_{mi}) - \int_\Omega d\Omega \, \rho_{mk} v_{mk} \frac{\partial (u_k \delta v_{mi})}{\partial x_i} \\
&= \int_\Gamma d\Gamma \, n_i u_k \rho_{mk} v_{mk} \delta v_{mi} - \int_\Omega d\Omega \, \rho_{mk} v_{mk} \delta v_{mi} \frac{\partial u_k}{\partial x_i} \\
&\quad - \int_\Omega d\Omega \, \rho_{mk} v_{mk} u_k \frac{\partial \delta v_{mi}}{\partial x_i} \\
&= \int_\Gamma d\Gamma \, \mathbf{n}(\mathbf{u} \cdot \rho_m \mathbf{v}_m) \cdot \delta \mathbf{v}_m - \int_\Omega d\Omega \, \nabla \mathbf{u} \cdot (\rho_m \mathbf{v}_m) \cdot \delta \mathbf{v}_m,
\end{aligned} \tag{A21}$$

$$\begin{aligned}
\int_\Omega d\Omega \, \mathbf{u} \cdot (\mathbf{v}_m \cdot \nabla) \rho_m \delta \mathbf{v}_m &= \int_\Omega d\Omega \, u_k v_{mi} \frac{\partial}{\partial x_i}(\rho_{mk} \delta v_{mk}) \\
&= \int_\Omega d\Omega \, \frac{\partial}{\partial x_i}(u_k v_{mi} \rho_{mk} \delta v_{mk}) - \int_\Omega d\Omega \, \rho_{mk} \delta v_{mk} \frac{\partial}{\partial x_i}(u_k v_{mi}) \\
&= \int_\Gamma d\Gamma \, n_i u_k v_{mi} \rho_{mk} \delta v_{mk} - \int_\Omega d\Omega \, \rho_{mk} \delta v_{mk} v_{mi} \frac{\partial u_k}{\partial x_i} \\
&\quad - \int_\Omega d\Omega \, \rho_{mk} \delta v_{mk} u_k \frac{\partial v_{mi}}{\partial x_i} \\
&= \int_\Gamma d\Gamma \, \mathbf{u}(\rho_m \mathbf{v}_m \cdot \mathbf{n}) \cdot \delta \mathbf{v}_m - \int_\Omega d\Omega \, (\rho_m \mathbf{v}_m \cdot \nabla) \mathbf{u} \cdot \delta \mathbf{v}_m
\end{aligned} \tag{A22}$$

and

$$\begin{aligned}
\int_\Omega d\Omega \, \mathbf{u} \cdot (\mathbf{v}_m \cdot \nabla)(\delta \alpha \mathbf{v}_m) &= \int_\Omega d\Omega \, u_k v_{mi} \frac{\partial}{\partial x_i}(\delta \alpha_k v_{mk}) \\
&= \int_\Omega d\Omega \, \frac{\partial}{\partial x_i}(u_k v_{mi} \delta \alpha_k v_{mk}) - \int_\Omega d\Omega \, \delta \alpha_k v_{mk} \frac{\partial}{\partial x_i}(u_k v_{mi}) \\
&= \int_\Gamma d\Gamma \, n_i u_k v_{mi} \delta \alpha_k v_{mk} - \int_\Omega d\Omega \, \delta \alpha_k v_{mk} v_{mi} \frac{\partial u_k}{\partial x_i} \\
&\quad - \int_\Omega d\Omega \, \delta \alpha_k v_{mk} u_k \frac{\partial v_{mi}}{\partial x_i} \\
&= \int_\Gamma d\Gamma \, \mathbf{u}(\mathbf{v}_m \cdot \mathbf{n}) \cdot \mathbf{v}_m \delta \alpha - \int_\Omega d\Omega \, (\mathbf{v}_m \cdot \nabla) \mathbf{u} \cdot \mathbf{v}_m \delta \alpha.
\end{aligned} \tag{A23}$$

Applying the tensor-vector identity [25], the divergence theorem and a property of the colon product, demonstrated below,

$$\begin{aligned}
\nabla \mathbf{u} : D(\delta \mathbf{v}_m) &= \nabla \mathbf{u} : \frac{1}{2}\left(\nabla \delta \mathbf{v}_m + (\nabla \delta \mathbf{v}_m)^T\right) \\
&= \frac{1}{2}\left(\nabla \mathbf{u} : \nabla \delta \mathbf{v}_m + \nabla \mathbf{u} : (\nabla \delta \mathbf{v}_m)^T\right) \\
&= \frac{1}{2}\left(\nabla \mathbf{u} : \nabla \delta \mathbf{v}_m + (\nabla \mathbf{u})^T : \nabla \delta \mathbf{v}_m\right) \\
&= \frac{1}{2}\left(\nabla \mathbf{u} + (\nabla \mathbf{u})^T\right) : \nabla \delta \mathbf{v}_m \\
&= D(\mathbf{u}) : \nabla \delta \mathbf{v}_m,
\end{aligned} \qquad (A24)$$

the term containing μ_m can be written as,

$$\begin{aligned}
\int_\Omega d\Omega \, \mathbf{u} \cdot \nabla \cdot (2\mu_m D(\delta \mathbf{v}_m)) &= \int_\Omega d\Omega \, \nabla \cdot (2\mu_m D(\delta \mathbf{v}_m) \cdot \mathbf{u}) - \int_\Omega d\Omega \, \nabla \mathbf{u} : 2\mu_m D(\delta \mathbf{v}_m) \\
&= \int_\Gamma d\Gamma \, 2\mu_m \mathbf{n} \cdot D(\delta \mathbf{v}_m) \cdot \mathbf{u} - \int_\Omega d\Omega \, 2\mu_m D(\mathbf{u}) : \nabla \delta \mathbf{v}_m \\
&= \int_\Gamma d\Gamma \, 2\mu_m \mathbf{n} \cdot D(\delta \mathbf{v}_m) \cdot \mathbf{u} - \int_\Omega d\Omega \, \nabla \cdot (2\mu_m D(\mathbf{u}) \cdot \delta \mathbf{v}_m) \\
&\quad + \int_\Omega d\Omega \, \nabla \cdot (2\mu_m D(\mathbf{u})) \cdot \delta \mathbf{v}_m \\
&= \int_\Gamma d\Gamma \, 2\mu_m \mathbf{n} \cdot D(\delta \mathbf{v}_m) \cdot \mathbf{u} - \int_\Gamma d\Gamma \, 2\mu_m \mathbf{n} \cdot D(\mathbf{u}) \cdot \delta \mathbf{v}_m \\
&\quad + \int_\Omega d\Omega \, \nabla \cdot (2\mu_m D(\mathbf{u})) \cdot \delta \mathbf{v}_m.
\end{aligned} \qquad (A25)$$

Similarly, the terms containing μ_d can be written as,

$$\begin{aligned}
\int_\Omega d\Omega \, \mathbf{u} \cdot \nabla \cdot (\delta_\alpha \mu_d D(\mathbf{v}_m)) &= \int_\Gamma d\Gamma \, \delta_\alpha \mu_d \mathbf{n} \cdot D(\mathbf{v}_m) \cdot \mathbf{u} - \int_\Gamma d\Gamma \, \delta_\alpha \mu_d \mathbf{n} \cdot D(\mathbf{u}) \cdot \mathbf{v}_m \\
&\quad + \int_\Omega d\Omega \, \nabla \cdot (\delta_\alpha \mu_d D(\mathbf{u})) \cdot \mathbf{v}_m
\end{aligned} \qquad (A26)$$

and

$$\begin{aligned}
\int_\Omega d\Omega \, \mathbf{u} \cdot \nabla \cdot (\delta_{\mathbf{v}_m} \mu_d D(\mathbf{v}_m)) &= \int_\Gamma d\Gamma \, \delta_{\mathbf{v}_m} \mu_d \mathbf{n} \cdot D(\mathbf{v}_m) \cdot \mathbf{u} - \int_\Gamma d\Gamma \, \delta_{\mathbf{v}_m} \mu_d \mathbf{n} \cdot D(\mathbf{u}) \cdot \mathbf{v}_m \\
&\quad + \int_\Omega d\Omega \, \nabla \cdot (\delta_{\mathbf{v}_m} \mu_d D(\mathbf{u})) \cdot \mathbf{v}_m.
\end{aligned} \qquad (A27)$$

Applying the product rule and divergence theorem, the remaining terms in Equation (11) can be written as,

$$\begin{aligned}
\int_\Omega d\Omega \, q \nabla \cdot (\rho_m \delta \mathbf{v}_m) &= \int_\Omega d\Omega \, \nabla \cdot (q \rho_m \delta \mathbf{v}_m) - \int_\Omega d\Omega \, \nabla q \cdot \rho_m \delta \mathbf{v}_m \\
&= \int_\Gamma d\Gamma \, q \rho_m \mathbf{n} \cdot \delta \mathbf{v}_m - \int_\Omega d\Omega \, \rho_m \nabla q \cdot \delta \mathbf{v}_m,
\end{aligned} \qquad (A28)$$

$$\begin{aligned}
\int_\Omega d\Omega \, \beta \nabla \cdot (\alpha \delta \mathbf{v}_m) &= \int_\Omega d\Omega \, \nabla \cdot (\beta \alpha \delta \mathbf{v}_m) - \int_\Omega d\Omega \, \nabla \beta \cdot \alpha \delta \mathbf{v}_m \\
&= \int_\Gamma d\Gamma \, \alpha \beta \mathbf{n} \cdot \delta \mathbf{v}_m - \int_\Omega d\Omega \, \alpha \nabla \beta \cdot \delta \mathbf{v}_m,
\end{aligned} \qquad (A29)$$

$$\int_\Omega d\Omega \, \mathbf{u} \cdot \nabla(\rho_m \delta p_m) = \int_\Omega d\Omega \, \nabla \cdot (\mathbf{u}\rho_m \delta p_m) - \int_\Omega d\Omega \, \nabla \cdot \mathbf{u}\rho_m \delta p_m$$
$$= \int_\Gamma d\Gamma \, \rho_m \mathbf{u} \cdot \mathbf{n} \delta p_m - \int_\Omega d\Omega \, \nabla \cdot \rho_m \mathbf{u} \delta p_m, \tag{A30}$$

$$\int_\Omega d\Omega \, \mathbf{u} \cdot \nabla(\delta \alpha p_m) = \int_\Omega d\Omega \, \nabla \cdot (\mathbf{u}\delta \alpha p_m) - \int_\Omega d\Omega \, \nabla \cdot \mathbf{u}\delta \alpha p_m$$
$$= \int_\Gamma d\Gamma \, \mathbf{u} \cdot \mathbf{n} \delta \alpha p_m - \int_\Omega d\Omega \, \nabla \cdot \mathbf{u}\delta \alpha p_m, \tag{A31}$$

$$\int_\Omega d\Omega \, \mathbf{u} \cdot \nabla \cdot \delta_\alpha(\alpha \rho_d \mathbf{v}_{dj} \mathbf{v}_{dj}) = \int_\Omega d\Omega \, \nabla \cdot \left(\mathbf{u} \delta_\alpha(\alpha \rho_d \mathbf{v}_{dj} \mathbf{v}_{dj}) \right)$$
$$- \int_\Omega d\Omega \, \nabla \cdot \mathbf{u} \delta_\alpha(\alpha \rho_d \mathbf{v}_{dj} \mathbf{v}_{dj}) \tag{A32}$$
$$= \int_\Gamma d\Gamma \, \mathbf{u} \cdot \mathbf{n} \delta_\alpha(\alpha \rho_d \mathbf{v}_{dj} \mathbf{v}_{dj})$$
$$- \int_\Omega d\Omega \, \nabla \cdot \mathbf{u} \delta_\alpha(\alpha \rho_d \mathbf{v}_{dj} \mathbf{v}_{dj}),$$

$$\int_\Omega d\Omega \, q \nabla \cdot (\delta \alpha \mathbf{v}_m) = \int_\Omega d\Omega \, \nabla \cdot (q \delta \alpha \mathbf{v}_m) - \int_\Omega d\Omega \, \nabla q \cdot (\delta \alpha \mathbf{v}_m)$$
$$= \int_\Gamma d\Gamma \, q \mathbf{v}_m \cdot \mathbf{n} \delta \alpha - \int_\Omega d\Omega \, (\mathbf{v}_m \cdot \nabla) q \delta \alpha, \tag{A33}$$

$$\int_\Omega d\Omega \, \beta \nabla \cdot (\delta \alpha \mathbf{v}_m) = \int_\Omega d\Omega \, \nabla \cdot (\beta \delta \alpha \mathbf{v}_m) - \int_\Omega d\Omega \, \nabla \beta \cdot (\delta \alpha \mathbf{v}_m)$$
$$= \int_\Gamma d\Gamma \, \beta \mathbf{v}_m \cdot \mathbf{n} \delta \alpha - \int_\Omega d\Omega \, (\mathbf{v}_m \cdot \nabla) \beta \delta \alpha, \tag{A34}$$

$$\int_\Omega d\Omega \, \beta \nabla \cdot \delta_\alpha(\alpha \mathbf{v}_{dj}) = \int_\Omega d\Omega \, \nabla \cdot \left(\beta \delta_\alpha(\alpha \mathbf{v}_{dj}) \right) - \int_\Omega d\Omega \, \nabla \beta \cdot \delta_\alpha(\alpha \mathbf{v}_{dj})$$
$$= \int_\Gamma d\Gamma \, \beta \delta_\alpha(\alpha \mathbf{v}_{dj}) \cdot \mathbf{n} - \int_\Omega d\Omega \, \delta_\alpha(\alpha \mathbf{v}_{dj}) \cdot \nabla \beta, \tag{A35}$$

$$\int_\Omega d\Omega \, \beta \frac{\mu_m^t}{\rho_c} \nabla \cdot \nabla \delta \alpha = \frac{\mu_m^t}{\rho_c} \left(\int_\Omega d\Omega \, \nabla \cdot (\beta \nabla \delta \alpha) - \int_\Omega d\Omega \, \nabla \beta \cdot \nabla \delta \alpha \right)$$
$$= \frac{\mu_m^t}{\rho_c} \Bigg(\int_\Gamma d\Gamma \, \beta \mathbf{n} \cdot \nabla \delta \alpha$$
$$- \int_\Omega d\Omega \, \nabla \cdot (\delta \alpha \nabla \beta) + \int_\Omega d\Omega \, \delta \alpha \nabla \cdot \nabla \beta \Bigg) \tag{A36}$$
$$= \frac{\mu_m^t}{\rho_c} \Bigg(\int_\Gamma d\Gamma \, \beta (\mathbf{n} \cdot \nabla) \delta \alpha$$
$$- \int_\Gamma d\Gamma \, \delta \alpha (\mathbf{n} \cdot \nabla) \beta + \int_\Omega d\Omega \, \delta \alpha \nabla \cdot \nabla \beta \Bigg),$$

$$\int_\Omega d\Omega \beta \frac{\mu_m^t}{\rho_c}\left(\frac{\rho_d}{\rho_c}-1\right)\nabla\cdot(\delta\alpha\nabla\alpha) = \frac{\mu_m^t}{\rho_c}\left(\frac{\rho_d}{\rho_c}-1\right)\left(\int_\Omega d\Omega\,\nabla\cdot(\beta\delta\alpha\nabla\alpha)\right.$$
$$\left.-\int_\Omega d\Omega\,\nabla\beta\cdot(\delta\alpha\nabla\alpha)\right) \quad (A37)$$
$$= \frac{\mu_m^t}{\rho_c}\left(\frac{\rho_d}{\rho_c}-1\right)\left(\int_\Gamma d\Gamma\,\beta\delta\alpha(\mathbf{n}\cdot\nabla)\alpha\right.$$
$$\left.-\int_\Omega d\Omega\,\delta\alpha\nabla\alpha\cdot\nabla\beta\right)$$

and

$$\int_\Omega d\Omega \beta \frac{\mu_m^t}{\rho_c}\left(\frac{\rho_d}{\rho_c}-1\right)\nabla\cdot(\alpha\nabla\delta\alpha) = \frac{\mu_m^t}{\rho_c}\left(\frac{\rho_d}{\rho_c}-1\right)\left(\int_\Omega d\Omega\,\nabla\cdot(\beta\alpha\nabla\delta\alpha)\right.$$
$$\left.-\int_\Omega d\Omega\,\nabla\beta\cdot(\alpha\nabla\delta\alpha)\right)$$
$$= \frac{\mu_m^t}{\rho_c}\left(\frac{\rho_d}{\rho_c}-1\right)\left(\int_\Gamma d\Gamma\,\beta\alpha\mathbf{n}\cdot\nabla\delta\alpha\right. \quad (A38)$$
$$\left.-\int_\Omega d\Omega\,\nabla\cdot(\alpha\delta\alpha\nabla\beta) + \int_\Omega d\Omega\,\delta\alpha\nabla\cdot(\alpha\nabla\beta)\right)$$
$$= \frac{\mu_m^t}{\rho_c}\left(\frac{\rho_d}{\rho_c}-1\right)\left(\int_\Gamma d\Gamma\,\alpha\beta(\mathbf{n}\cdot\nabla)\delta\alpha\right.$$
$$\left.-\int_\Gamma d\Gamma\,\alpha\delta\alpha(\mathbf{n}\cdot\nabla)\beta + \int_\Omega d\Omega\,\delta\alpha\nabla\cdot(\alpha\nabla\beta)\right).$$

Equation (11) can now be reformulated and rearranged as Equation (13).

Appendix E. Derivation of Equation (36c)

Decomposing the dispersed-phase velocity into the mixture and dispersed-phase diffusion velocities, Equation (34) can be rewritten as,

$$J_\Gamma = \alpha\rho_d(\mathbf{v}_m + \mathbf{v}_{dm})\cdot\mathbf{n}$$
$$= \alpha\rho_d\left(\mathbf{v}_m + \frac{\rho_c}{\rho_m}\mathbf{v}_{dj}\right)\cdot\mathbf{n}, \quad (A39)$$

where $\mathbf{v}_{dj}$ is defined in terms of $\mathbf{v}_{dm}$, the dispersed-phase velocity relative to the mixture velocity, as $\mathbf{v}_{dj} = \frac{\rho_m}{\rho_c}\mathbf{v}_{dm}$. Substituting Equation (A6) into Equation (A39) and rewriting the parentheses as binomial expansions,

$$J_\Gamma = \alpha\rho_d\left(\mathbf{v}_m + (1-\alpha)^{-1}\left(1 + \frac{\alpha}{1-\alpha}\frac{\rho_d}{\rho_c}\right)^{-1}\mathbf{v}_{dj}\right)\cdot\mathbf{n} \quad (A40)$$

$$= \alpha\rho_d\left(\mathbf{v}_m + (1+\alpha+\cdots)\left(1 - \frac{\alpha}{1-\alpha}\frac{\rho_d}{\rho_c} + \cdots\right)\mathbf{v}_{dj}\right)\cdot\mathbf{n} \quad (A41)$$

$$= \alpha\rho_d\left(\mathbf{v}_m + \left(1 + \alpha\left(1 - \frac{1}{1-\alpha}\frac{\rho_d}{\rho_c}\right) + \cdots\right)\mathbf{v}_{dj}\right)\cdot\mathbf{n}. \quad (A42)$$

As $\alpha \ll 1$ and $\rho_d \approx 2\rho_c \implies \left|1 - \frac{1}{1-\alpha}\frac{\rho_d}{\rho_c}\right| < 2$ and ignoring terms containing squared and higher powers of α,

$$J_\Gamma \approx \alpha\rho_d(\mathbf{v}_m + \mathbf{v}_{dj})\cdot\mathbf{n}. \quad (A43)$$

Applying the product rule,

$$\begin{aligned}\frac{\partial J_\Gamma}{\partial \alpha} &= \rho_d(\mathbf{v}_m + \mathbf{v}_{dj}) \cdot \mathbf{n} + \alpha \rho_d \frac{\partial \mathbf{v}_{dj}}{\partial \alpha} \cdot \mathbf{n} \\ &= \rho_d \left(\mathbf{v}_m + \mathbf{v}_{dj} + \alpha \frac{\partial \mathbf{v}_{dj}}{\partial \alpha} \right) \cdot \mathbf{n}.\end{aligned} \quad (A44)$$

References

1. Soto, O.; Löhner, R.; Yang, C. An adjoint-based design methodology for CFD problems. *Int. J. Num. Meth. Heat Fluid Flow* **2004**, *14*, 734–759. [CrossRef]
2. Giles, M.; Pierce, N. An introduction to the adjoint approach to design. *Flow Turb. Combust.* **2000**, *65*, 393–415. [CrossRef]
3. Othmer, C. Adjoint methods for car aerodynamics. *J. Math. Ind.* **2014**, *4*, 1–23. [CrossRef]
4. Karpouzas, G.K.; Papoutsis-Kiachagias, E.M.; Schumacher, T.; de Villiers, E.; Giannakouglou, K.C.; Othmer, C. Adjoint Optimisation for Vehicle External Aerodynamics. *Int. J. Automot. Eng.* **2016**, *7*, 1–7.
5. Papoutsis-Kiachagias, E.M.; Asouti, V.G.; Giannakoglou, K.C.; Gkagkas, K.; Shimokawa, S.; Itakura, E. Multi-Point Aerodynamic Shape Optimization of Cars Based on Continuous Adjoint. *Struct. Multidiscip. Optim.* **2019**, *59*, 675–694. [CrossRef]
6. Reuther, J.; Jameson, A.; Farmer, J.; Martinelli, L.; Saunders, D. *Aerodynamic Shape Optimization of Complex Aircraft Configurations Via an Adjoint Formulation*; Paper 94; AIAA: Reston, VA, USA, 1996.
7. Kroll, N.; Gauger, N.; Brezillon, J.; Dwight, R.; Fazzolari, A.; Vollmer, D.; Becker, K.; Barnewitz, H.; Schulz, V.; Hazra, S. Flow simulation and shape optimization for aircraft design. *J. Comput. Appl. Math.* **2007**, *203*, 397–411. [CrossRef]
8. Campobasso, M.; Duta, M.; Giles, M. Adjoint calculation of sensitivities of turbomachinery objective functions. *J. Propul. Power* **2003**, *19*, 693–703. [CrossRef]
9. Wang, D.X.; He, L. Adjoint aerodynamic design optimization for blades in multistage turbomachines; part I: methodology and verification. *J. Turbomach.* **2010**, *132*, 021011. [CrossRef]
10. Wang, D.; He, L.; Li, Y.; Wells, R. Adjoint aerodynamic design optimization for blades in multistage turbomachines–part II: validation and application. *J. Turbomach.* **2010**, *132*, 021012. [CrossRef]
11. *Shape Optimization for Aerodynamic Efficiency using Adjoint Methods*; ANSYS White Paper; ANSYS: Canonsburg, PA, USA, 2016.
12. Alexias, P.; Giannakoglou, K.C. Shape Optimisation of a Two-Fluid Mixing Device using Continuous Adjoint. *Fluids* **2020**, *5*, 11. [CrossRef]
13. Martin, M.J.; Andres, E.; Lozano, C.; Valero, E. Volumetric B-splines shape parametrization for aerodynamic shape design. *Aerosp. Sci. Technol.* **2014**, *37*, 26–36. [CrossRef]
14. Jakobsson, S.; Amoignon, O. Mesh deformation using radial basis functions for gradient-based aerodynamic shape optimisation. *Comput. Fluids* **2007**, *36*, 1119–1136. [CrossRef]
15. Kapellos, C.; Alexias, P.; De Villiers, E. The adjoint mehod for automotive optimisation using a sphericity based morpher. In Proceedings of the International Association for the Engineering Modelling, Analysis and Simulation Adjoint CFD Seminar (NAFEMS Adjoint CFD Seminar), Wiesbaden, Germany, 23–24 October 2016.
16. Drew, D.A. Mathematical Modeling of Two-Phase Flow. *Ann. Rev. Fluid Mech.* **1983**, *15*, 261–291. [CrossRef]
17. Balachandar, S.; Eaton, J. Turbulent Dispersed Multiphase Flow. *Ann. Rev. Fluid Mech.* **2010**, *42*, 111–133. [CrossRef]
18. Brennan, D. The Numerical Simulation of Two-Phase Flows in Settling Tanks. Ph.D. Thesis, Imperial College London, London, UK, 2001.
19. Othmer, C. A continuous adjoint formulation for the computation of topological and surface sensitivities of ducted flows. *Int. J. Numer. Methods Fluids* **2008**, *58*, 861–877. [CrossRef]
20. Kavvadias, I.S.; Papoutsis-Kiachagias, E.M.; Dimitrakopoulos, G.; Giannakoglou, K.C. The continuous adjoint approach to the k-ω SST turbulence model with applications in shape optimisation. *Eng. Optim.* **2015**, *47*, 1523–1542. [CrossRef]

21. Schramm, M.; Stoevesandt, B.; Peinke, J. Optimisation of Airfoils using the Adjoint Approach and the Influence of Adjoint Turbulent Viscosity. *Computation* **2018**, *6*, 23. [CrossRef]
22. Ubbink, O. Numerical prediction of two fluid systems with sharp interfaces. Ph.D. Thesis, Imperial College London, London, UK, 1997.
23. Dahl, C. Numerical Modelling of Flow and Settling in Secondary Settling Tanks. Ph.D Thesis, Aalborg University, Aalborg, Denmark, 1993.
24. Takacs, I.; Patry, G.G.; Nolasco, D. A dynamic model of the clarification-thickening process. *Water Res.* **1991**, *25*, 1263–1271. [CrossRef]
25. Clarke, D.A. A Primer on Tensor Calculus. Department of Physics, Saint Mary's University, Halifax, NS, Canada. Unpublished manuscript, 2011.

© 2020 by the authors. Licensee MDPI, Basel, Switzerland. This article is an open access article distributed under the terms and conditions of the Creative Commons Attribution (CC BY) license (http://creativecommons.org/licenses/by/4.0/).

Article

Cloud-Based CAD Parametrization for Design Space Exploration and Design Optimization in Numerical Simulations

Joel Guerrero [1,*], Luca Mantelli [2] and Sahrish B. Naqvi [1]

[1] DICCA, Department of Civil, Chemical, and Environmental Engineering, University of Genoa, Via Montallegro 1, 16145 Genoa, Italy; 4457071@studenti.unige.it

[2] DIME, Department of Mechanical, Energy, and Transportation Engineering, Thermochemical Power Group (TPG), University of Genoa. Via Montallegro 1, 16145 Genoa, Italy; luca.mantelli@edu.unige.it

* Correspondence: joel.guerrero@unige.it

Received: 2 February 2020; Accepted: 13 March 2020; Published: 18 March 2020

Abstract: In this manuscript, an automated framework dedicated to design space exploration and design optimization studies is presented. The framework integrates a set of numerical simulation, computer-aided design, numerical optimization, and data analytics tools using scripting capabilities. The tools used are open-source and freeware, and can be deployed on any platform. The main feature of the proposed methodology is the use of a cloud-based parametrical computer-aided design application, which allows the user to change any parametric variable defined in the solid model. We demonstrate the capabilities and flexibility of the framework using computational fluid dynamics applications; however, the same workflow can be used with any numerical simulation tool (e.g., a structural solver or a spread-sheet) that is able to interact via a command-line interface or using scripting languages. We conduct design space exploration and design optimization studies using quantitative and qualitative metrics, and, to reduce the high computing times and computational resources intrinsic to these kinds of studies, concurrent simulations and surrogate-based optimization are used.

Keywords: CFD; numerical optimization; CAD parametrization; cloud-based; design space exploration; SSIM

1. Introduction

Consumer demand, government regulations, competitiveness, globalization, better educated end-users, environmental concerns, market differentiation, social media trends, and even influencers, they are all driving products manufacturers and industry to reduce production expenditures and final cost of goods, and at the same time improving the quality and reliability of the products with the lowest environmental impact. To reach these goals and to develop revolutionary products, the manufacturing sector is relying more on virtual prototypes, computer simulations, and design optimization.

Computational fluid dynamics (CFD), computational structural dynamics (CSD), computer-aided manufacturing (CAM), computer-aided design (CAD), multi-physics simulations, digital twins, the internet-of-things (IoT) and the cloud, are among many of the tools increasingly being used to simulate and certify products by analysis and simulation before going into production and commercialization. Even before reaching the market, modern products have undergone some kind of heuristic or methodological optimization. Though the optimization might take different forms in different fields (e.g., finance, health, construction, operations, manufacturing, transportation, construction, engineering design, sales, public services, mail, and so on), the ultimate goal is always getting the best out of something under given circumstances, either by minimizing, maximizing, equalizing, or zeroing a quantity of interest (QoI).

Product optimization can be undertaken in two different ways, by using design space exploration (DSE) or by using design optimization (DO). Even a combination of both methodologies is possible. In DSE, we simply explore the design space in a methodological way, and while doing so, we extract knowledge. DSE is the process of discovering, expanding, evolving, and navigating the design space to extract knowledge to support better decision making [1]. It is not difficult to recognize that in DSE, we are not converging to an optimal value, we are only exploring the design space, but in doing so, we are gathering valuable information about the global behavior, and this information can be used to get a better design. Moreover, this knowledge can also be used to conduct surrogate-based optimization (SBO) studies. The SBO method consists of constructing a mathematical model (also known as a surrogate, response surface, meta-model, emulator) from a limited number of observations (CFD simulations, physical experiments, or any quantifiable metric) [2–5]. After building the surrogate, it can be explored and exploited. Conducting the optimization at the surrogate level is orders of magnitude faster than working at the high fidelity level [2].

Design optimization strategies, on the other hand, consist on formulating an optimization problem and converging to the optimal design. Here, it is assumed that the problem can be formulated before the search and convergence begin. A typical optimization problem can be formulated as follows,

$$\text{Find} \quad \mathbf{X} = \begin{Bmatrix} x_1 \\ x_2 \\ \vdots \\ x_n \end{Bmatrix} \tag{1}$$

which minimizes, maximizes, equalizes, or zeroed,

$$f_j(\mathbf{X}), \quad j = 1, 2, \cdots, q \tag{2}$$

subject to design constraints (linear and non-linear),

$$\begin{aligned} g_j(\mathbf{X}) &\leq 0, \quad j = 1, 2, \cdots, m \\ l_j(\mathbf{X}) &= 0, \quad j = 1, 2, \cdots, p \end{aligned} \tag{3}$$

and variables bounds,

$$x_i^{lb} \leq x_i \leq x_i^{ub} \quad j = 1, 2, \cdots, n \tag{4}$$

where $\mathbf{X}$ is a n-dimensional vector called the design vector, $f_j(\mathbf{X})$ is the objective function or QoI, $g_j(\mathbf{X})$ are the inequality constraints, $l_j(\mathbf{X})$ are the equality constraints, and x_i^{lb} and x_i^{ub} are the variables lower and upper bounds, respectively. To find the optimal value we can use gradient-based methods or derivative-free methods [5–10]. Also, the problem formulation can be single-objective (one QoI to be optimized) or multi-objective (more than one QoI to be optimized simultaneously). Things can get even more complicated, as in some cases we might need to deal with design optimization problems incorporating many disciplines (e.g., aerodynamics, propulsion, structures, and performance). In this case, we say we are dealing with a multi-disciplinary design optimization problem (MDO) [11–16]. MDO allows designers and engineers to incorporate all relevant disciplines simultaneously. The optimum of the simultaneous problem is superior to the design found by optimizing each discipline sequentially since it can exploit the interactions between the disciplines. However, including all disciplines simultaneously significantly increases the complexity of the problem [7].

The field we are concerned with in this manuscript is that of engineering design; nevertheless, this by no means limits the range of applicability of the current work; it simply reflects the authors' interests and fields of expertise.

In engineering design, we are often interested in optimizing the geometry. To do so, two approaches are available, direct modeling and parametric modeling. In direct modeling, we modify the geometry by pushing and pulling points, lines, and surfaces (like working with clay). This gives designers and engineers a lot of flexibility when it comes to shape the geometry; however, in the process of doing so, we give up geometry parametrization in favor of creating organic shapes that might be difficult to manufacture. In parametric modeling, the user defines relationships, constraints, parametric variables, and configurations when creating the solid model. Then, by changing these variables, the user can easily create endless variations on the original geometry with complete control and millimetric precision.

However, when conducting fully automatic DSE or DO studies, introducing the CAD tools is not very straightforward. Most of the times the CAD applications are not compatible with the operating system (OS) where the numerical simulations are being performed (usually Unix-like OS), or simply, it is not possible to connect the optimization loop with the CAD tool due to the fact that the user can only interact with it using a graphical user interface (GUI), which cannot be used in an automatic optimization loop driven by a command-line interface (CLI).

To overcome this problem, many commercial simulation frameworks are adding a monolithic design environment to integrate all the applications needed to conduct design space exploration and design optimization studies, namely CAD, multi-physics solver, optimizer, and post-processing. While commercial frameworks have proven to be reliable, they come with a price tag that often is unreachable by small and medium-sized enterprises (SMEs), hobbyists, researchers or personal users. Hereafter, we propose the integration of open-source and freeware tools to conduct DSE and DO studies.

To perform the numerical simulations, we use the multi-physics solver OpenFOAM (version 7.0) [17,18] or the programming language Python. The optimization algorithms and the code coupling interface is provided via the Dakota library [19,20] (version 6.10). All the real-time data analytics, quantitative and qualitative post-processing, and data analytics are performed using Python, VTK [21], and bash scripting. Finally, to create and modify the geometry we use Onshape [22], which is a cloud-based parametric CAD and product development application. Onshape's application programming interface (API) is open-source; therefore, it can be deployed in any platform with an internet connection. The API is implemented in Python, and the calls to Onshape's server are done using RESTful requests. Onshape offers two subscription plans, a pay-up plan and a free one. Both subscriptions plans have the same professional capabilities, the only difference is the level of product support offered and the access to enterprise options.

The purpose of this manuscript is two-fold. First, we want to use the cloud to support CAD parametrization in DSE or DO design loops, which undoubtedly will give users enormous flexibility as the CAD application does not need to be installed locally, and there is no need for a monolithic CAD/Simulation software integration. Secondly, we want to deploy fully automatic, fault-tolerant, and scalable engineering design loops using in-house computational resources, the cloud, or HPC centers; and everything based on open-source and freeware tools. We hope that this contribution will offer guidelines to designers and engineers working with design optimization and design space exploration, will help them at implementing their own optimization loops, and to some extent, it will help to address some of the findings and recommendations listed in the NASA contractor report *"CFD Vision 2030 Study: A Path to Revolutionary Computational Aerosciences"* [23], where it is stated the following: *"Included in this desired set of capabilities is a vision for how CFD in 2030 will be used: a vision of the interaction between the engineer/scientist, the CFD software itself, its framework and all the ancillary software dependencies (databases, modules, visualization, etc.), and the associated HPC environment. A single engineer/scientist must be able to conceive, create, analyze, and interpret a large ensemble of related simulations in a time-critical period (e.g., 24 h), without individually managing each simulation, to a pre-specified level of accuracy"*.

The rest of the manuscript is organized as follows. Section 2 gives an overview of the methodology used. In Section 3 we describe the numerical experiments carried out to demonstrate the usability and flexibility of the framework. Finally, in Section 4 we present the conclusions and future perspectives.

2. Description of the Workflow—Methodology

In Figure 1, we illustrate a graphical summary of the methodology used in this work. The engineering design loop starts with a fully parametrized geometry, then new candidates are generated by changing the parametrical variables. It is important to stress that our starting point is the parametrical variables and not the solid model; that is, we are allowed to start from any possible geometry that can be generated using the parametrical variables. Hereafter, we use Onshape [22] as solid modeler, which is a cloud-based CAD application. The fact that Onshape is cloud-based gives us the flexibility to deploy the framework in any platform without the need to install the application. The only requirement is to have a working internet connection.

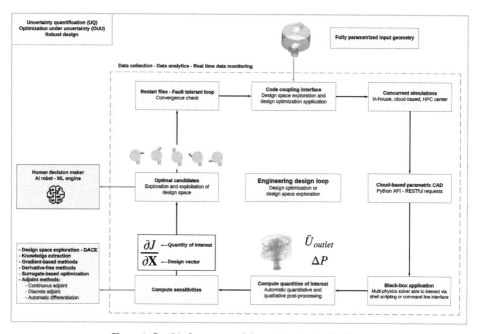

Figure 1. Graphical summary of the engineering design loop.

The whole workflow is controlled by the library Dakota [19,20], which serves as the numerical optimizer and code coupling interface tool. The Dakota library provides a flexible and extensible interface between simulation codes and iterative analysis methods. The library is software agnostic, in the sense that it can interface any application that is able to parse input/output files via a CLI. The library also has extensive design optimization and design space exploration capabilities. It comes with many gradient-based methods and derivative-free methods for design optimization. It also contains many design and analysis of computer experiments (DACE) methods to conduct design space exploration studies. And to obtain faster turn-around times, Dakota supports concurrent function evaluations.

The engineering design loop illustrated in Figure 1 is orchestrated by using Dakota's configuration input file. In this input file, all the steps to follow in the engineering design loop are defined. As previously stated, the only requirement is that the applications involved in the loop can interact via the CLI. In references [3,24–33], few examples using Dakota to control complex engineering design loops are discussed. However, none of them addressed the use of a fully parametric cloud-based CAD tool to generate the solid geometry or the use of the cloud to deploy the loop.

After defining Dakota's configuration file, the engineering design loop can be launched sequentially or concurrently using local resources, on the cloud, or remotely in an HPC center. All the tools involved in the loop are black-box applications that are connected using Dakota. An essential step of every optimization loop is that a QoI must be provided to compute the sensitivities; this is also controlled using Dakota's configuration file. This step is critical and is the user's responsibility to define all the quantities of interests to monitor. After computing the QoI, Dakota will compute the sensitivities using the method selected by the user. With Dakota, the user is not obliged to use the optimization and space exploration methods implemented on it; one can easily interface Dakota with a third-party optimization library.

At this point, we can rely on a human decision-maker or a machine learning engine to pick up the best design or set of optimal solutions. During the whole process, data is collected and monitored in real time. Dakota also offers restart capabilities, so in the event of an unexpected failure of the system (hardware or software), the user can restart from a previously saved state.

In this work, we use the design loop illustrated in Figure 1 for DO and DSE studies. In DO, the user starts from an initial design or guess, and the optimization algorithm will make it slightly better, i.e., in DO we are making sub-optimal guesses incrementally better. This by no means is negative, and the chances are that the results are a substantial improvement over the initial guess. In essence, DO is an iterative-converging process that requires a starting point (or a set of points) and a set of constraints. On the other hand, in DSE we do not need to define an initial guess or a set of constraints (except for the bounds of the design space). We generate new solutions sequentially or concurrently that might be better or worse than a baseline, but in the process of doing so, we are exploring and exploiting the design space. DSE gives more information to engineers than DO, and this information can be used for decision making, knowledge extraction, and anomalies detection. All the information gathered during the design loop can also be used to construct reduced-order models, surrogate models, or to interrogate the data using exploratory data analysis and machine learning techniques.

3. Numerical Experiments

3.1. Cylinder Optimization Problem—Minimum Surface and Fixed Volume

This problem is also known as the soda can optimization problem. We aim at finding the optimal dimensions of a right cylinder that minimize the total surface area of the cylinder, which holds a given volume. This problem can be formulated as follows,

$$\text{minimize } S_{tot} \tag{5}$$

subject to,

$$V = 355 \, \text{cm}^3 \\ 0 < r, h < \infty \tag{6}$$

where

$$S_{tot} = 2\pi r^2 + 2\pi r h \\ V = \pi r^2 h \tag{7}$$

in Equations (5)–(7), S_{tot} is the cylinder's total surface, V its volume, r its radius, and h its height. The solution to this problem is the following,

$$r = 3.837 \text{ cm}$$
$$h = 7.675 \text{ cm} \qquad (8)$$
$$S_{min} = 277.54 \text{ cm}^2$$

This is a classic problem that is frequently posed to first-year calculus students. Therefore, we will not go into details on how to find the analytical solution (Equation (8)). Instead, we will use this case to illustrate how the cloud-based design loop works.

In Figure 2, we illustrate the general workflow. In steps 1–2, we define all the configuration variables and measurements (e.g., area, volume, length, and so on). In these steps, we also check that we are obtaining the desired output by changing manually the parametrical variables. In Figure 3, we show the screen-shot of how this case was setup in Onshape (the document is available at the following link https://cad.onshape.com/documents/448249f25f37397d1823feb6/w/33bca1cf858efd73dc35ab4f/e/2ec99afd57f87dd94045affd); in the figure, it can be observed that all the configurations, bounds, and measurements have been defined. All these variables can be accessed or modified using Onshape's Python API (https://github.com/onshape-public/apikey/tree/master/python). In step 3 we proceed to test the connection with Onshape's server, this is illustrated in Figure 4. In the figure, we use the API client to encode the changes to the model configurations and evaluation of the measurements. Then, using OAuth authentication, a RESTful request is sent to Onshape's server, which sends a response back to the client. The response can be the new geometry or the evaluation of the volume of the new solid model. After testing the configurations and communication with Onshape's server, we proceed to define the problem in Dakota's configuration file and to create any additional scripts needed to parse input/output files (step 4). This step includes choosing the optimization or space exploration method and defining the bounds, constraints, and objective functions. At this point, we can proceed to deploy the case sequentially or concurrently using local resources, the cloud, or HPC center resources (step 5). Finally, in step 6, we can visualize the optimal solid model. Additionally, we can use exploratory data analysis to study the collected data. During the whole process, restart files are generated, and data is monitored in real time.

In Listing 1, we show an excerpt of the Python code used to change the configuration variables. In the listing, the keywords **height_to_update**, **dia1_to_update**, and **dia2_to_update** are the parametric variables, and each one was defined in the Onshape document. Their values are substituted automatically by Dakota, and their bounds are defined in Dakota's configuration file. The function **part_studio_stl_conf** is responsible for exporting the geometry using the current values of the configuration variables (in this case the geometry is exported in STL format but any supported CAD exchange format can be used). The exported geometry is then used with the black-box solver. The **did**, **wid**, and **eid** keywords in Listing 1 are referred to the document id, workspace id, and element id of the Onshape document (refer to Figure 3). In Listing 2, we show an excerpt of the Python code used to evaluate the measurements (the structure is similar to that of Listing 1). In the listing, the line of code "**function(context, queries) return getVariable(context, 'volume');**" evaluates the measurement, as defined in the Onshape document. In this case, we are evaluating the volume of the solid model. As for the configuration variables, all the measurements need to be defined in the Onshape document. In the listing, the function **featurescript_conf** takes the configuration values and the measurement function definition and gives as output the evaluation of the measurement for the given configuration. For the interested reader, the working case with all the scripts can be downloaded at this link (https://github.com/joelguerrero/cloud-based-cad-paper/tree/master/soda_can/). These scripts can be used as a starting point for more complex cases. It is worth mentioning that the Python API works with Python 2 (2.7.9+).

Let us discuss the outcome of a DO study using a gradient-based method (method of feasible directions or MFD [34,35] with numerical gradients computed using forward differences). As we are optimizing a right cylinder, we set the diameters of the top and bottom surfaces to the same value, we also started to iterate from two different initial conditions. In Table 1, we show the outcome of this study. As can be observed, in both situations we arrived at the optimal value, and any deviation from the analytical solution is due to numerical precision and convergence tolerances. It is also interesting to note that depending on the starting conditions, different convergence rates can be achieved. The closer we are to the optimal solution, the faster the convergence will be. This put in evidence that the formulation of an optimization problem using gradient-based methods requires certain knowledge of the behavior of the design space; otherwise, the convergence rate to the optimal value will be slow.

Listing 1. Excerpt of the Python code used to setup the parametric configuration variables.

```
configuration = {
'units': 'meter',
'scale': 1.0,
'configuration' :
        'height={[height_to_update]}+m;'
        'dia1={[dia1_to_update]}+m;'
        'dia2={[dia2_to_update]}+m'
}

stl = c.part_studio_stl_conf(did, wid, eid, configuration)
```

Listing 2. Excerpt of the Python code used to evaluate the measurements.

```
body_feature = {
                "script" :
                "function(context, queries) {return getVariable(context, 'volume');}"
                }

configuration = {
'units': 'meter',
'scale': 1.0,
'configuration' :
        'height={[height_to_update]}+m;'
        'dia1={[dia1_to_update]}+m;'
        'dia2={[dia1_to_update]}+m'
}

out = c.featurescript_conf(did, wid, eid, body_feature, configuration)
```

During the DO study, we also used a derivative-free method (mesh adaptive direct search algorithm or MADS [36]), which also converged to the optimal solution but with a slow convergence rate, as shown in Table 2. As a side note, even if the derivative-free method exhibited a slow convergence rate, it was faster than the gradient-based method with a poor guess of the starting point (MFD-2 in Table 1). In general, derivative-free methods do not require the definition of the starting point, and they are insensitive to numerical noise.

Table 1. Outcome of the optimization study using a gradient-based method (MFD [34,35]).

	MFD-1	MFD-2	Analytical Solution
Starting point-Height (height_to_update)-cm	4	2	-
Starting point-Diameter (dia1_to_update)-cm	8	12	-
Optimal value-Height (height_to_update)-cm	7.617	7.607	7.675
Optimal value-Diameter (dia1_to_update)-cm	7.692	7.697	7.674
QoI (S_{tot})-cm^2	277.026	277.027	277.54
Non-linear constraint (Volume)-cm^3	354.001	354.000	354.98
Function evaluations	88	405	-

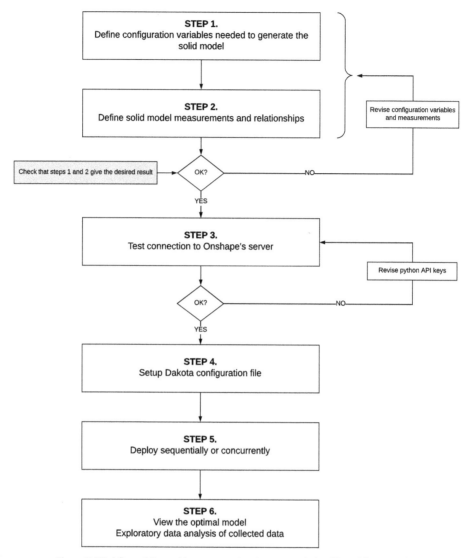

Figure 2. Workflow of the problem setup using the proposed cloud-based framework.

Table 2. Outcome comparison of the gradient-based method (MFD [34,35]) and the derivative-free method (MADS [36]). In the table, MFD refers to the gradient-based method (same as MFD-1 in Table 1), and MADS refers to the derivative-free method.

	MFD	MADS	Analytical Solution
Optimal value-Height (height_to_update)-cm	7.617	7.699	7.675
Optimal value-Diameter (dia1_to_update)-cm	7.692	7.655	7.674
QoI (S_{tot})-cm^2	277.026	277.236	277.54
Non-linear constraint (Volume)-cm^3	354.001	354.406	354.98
Function evaluations	88	256	-

In Table 3, we compare the results of the same DO study but this time using two and three design variables. Again, we obtain results close to the analytical one, and surprisingly, the convergence rate of both cases was similar. The main reason for the similarity of the convergence rate is that the starting points of the design variables are close to the optimal value. This evidence the importance of choosing good starting points to get a good convergence rate; gradient-based methods can be very sensitive to this choice. Regarding the case setup, the main difference is that we need to add additional scripts to compute the area of the top and bottom surfaces of the cylinder, independently.

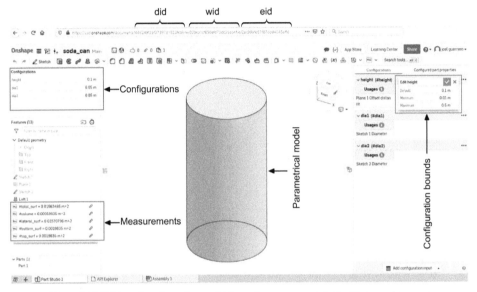

Figure 3. Definition of configuration variables and measurements in the Onshape document.

Let us run the same case using a design space exploration method. We remind the readers that when using DSE, we are not explicitly converging to an optimal solution; we are just exploring the design space. Then, the outcome of this study can be used for knowledge extraction, anomalies detection, or to construct a surrogate model. To conduct this DSE study, we used a full-factorial experiment with 21 experiments equally spaced for each design variable (for a total of 441 observations). In Figure 5, we show one of the many plots that can be used to visualize the data coming from DSE studies [3,37]. This plot is called scatter plot matrix, and in one single illustration, it shows the correlation information, the data distribution (using histograms and scatter plots), and regression models of the responses of the QoI.

By conducting a quick inspection of the scatter plot matrix displayed in Figure 5, we can demonstrate that the data is distributed uniformly in the design space (meaning that the sampling plan is unbiased), and this is demonstrated in the diagonal of the plot (the plots corresponding to the design variables). By looking at the scatter plot of the experiments (lower triangular part of the matrix), we see the distribution of the data in the design space. If, at this point, we detect regions in the design space that remain unexplored, we can add new training points to cover those areas. In the case of outliers (anomalies), we can remove them from the dataset with no significant inconvenience. However, we should be aware that outliers are telling us something, so it is a good idea to investigate the cause and effect of the outliers. In the upper triangular part of the plot, the correlation information is shown (Spearman correlation in this case). This information tells us how correlated the data is. For example, and by looking at the last row of the plot that shows the response of the QoI, if we note here a strong correlation between two variables, it is clear that these variables cannot be excluded from the study. As can be seen, this simple plot can be used to gather a deep understanding of the problem.

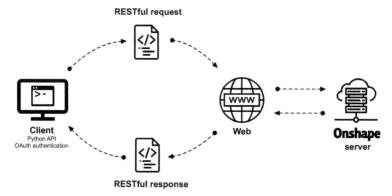

Figure 4. Onshape's cloud-based client-server communication using RESTful API. The client communicates with Onshape's server using Python API keys and OAuth authentication.

Table 3. Outcome of the optimization study using a gradient-based method (MFD [34,35]). In the table, MFD-2DV refers to the case with two design variables. MFD-3DV refers to the case with three design variables. The case MFD-2DV uses the same diameter for the top and bottom surfaces.

	MFD-2DV	MFD-3DV	Analytical Solution
Starting point-Height (height_to_update)-cm	4	4	-
Starting point-Diameter 1 (dia1_to_update)-cm	8	8	-
Starting point-Diameter 2 (dia2_to_update)-cm	-	5	-
Optimal value-Height (height_to_update)-cm	7.617	7.648	7.675
Optimal value-Diameter 1 (dia1_to_update)-cm	7.692	7.686	7.674
Optimal value-Diameter 2 (dia2_to_update)-cm	-	7.666	-
QoI (S_{tot})-cm^2	277.026	277.026	277.54
Non-linear constraint (Volume)-cm^3	354.001	354.004	354.98
Function evaluations	88	114	-

The data gathered from the DSE study can also be used to construct a meta-model, and then conduct the optimization at the surrogate level. In Figure 6, we illustrate the response surface, which was constructed using Kriging interpolation (universal Kriging). The implementation details of the method can be found in references [2,4,20,38–42]. To conduct the optimization at the surrogate level, we used the MFD gradient-based method (method of feasible directions [34,35] with analytical gradients). However, any optimization method (gradient-based or derivative-free) can be used as working at the surrogate level is inexpensive; we do not need to perform high-fidelity function evaluations.

In Figures 5 and 6, we plot a two-variable design space. In general, a design space will be n-dimensional, where n is the number of design variables of which the objective is a function. We deliberately used a two-variable design space to help visualize the response surface, the design space, and the various concepts related to DO and DSE. For completeness, we extended this problem to three design variables, and we obtained similar results by using the same methodology. We want to point out that all the results discussed in this section were obtained using Python scripting as black-box solver, and the volume and surfaces were computed using Onshape's API.

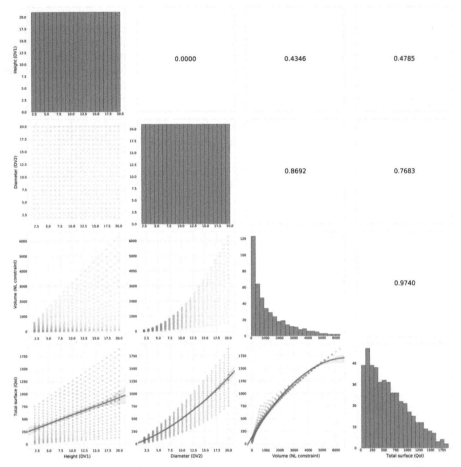

Figure 5. Scatter plot matrix of the cylinder optimization case using two design variables. In the upper triangular part of the plot, the Spearman correlation is shown. In the diagonal of the matrix, the histograms showing the data distribution are displayed. In the lower triangular part of the matrix, the data distribution is shown using scatter plots. In the last row of the matrix plot, the response of the QoI in function of the design variables and the non-linear (NL) constraint is illustrated, together with a quadratic regression model.

We would like to highlight that the optimized can dimensions presented in this section significantly differ from actual soda cans. We should ask ourselves, is the shape of this soda can truly optimal? From a mathematical point of view, yes. However, from a point of view of going through the whole process of manufacturing the can, is not. This simple example shows that optimization is very subjective. Sometimes manufacturers are trying to optimize something a little bit more abstract, like, how the can is manufactured, packing factor, opening mechanism, customer satisfaction, aluminum cost, and these abstract questions are better answered using design space exploration and by visualizing and interacting with the results in real time, as is possible to do by using the proposed cloud-based engineering design framework.

To close the discussion of this introductory case, we would like to reiterate that the optimization loop implemented is fault-tolerant, so in the event of hardware or software failure, the optimization task can be restarted from the last saved state. During the design loop, all the data is made available immediately to the user, including the geometry, even when running multiple simulations at the same time. Moreover, the data is monitored in real time; therefore, anomalies and trends can be detected in real time, and corrections/decisions can be taken. Finally, when it comes to engineering design studies, DO will converge to the optimal value, but formulating the problem requires some knowledge about the design space. Also, DO does not give valuable information about the global behavior of the QoI. Design space exploration, on the other hand, provides a lot of information about the design space without converging to the optimal value. Still, these studies might be expensive to conduct due to the high number of function evaluations often required to construct a reliable estimator. An added benefit of DSE is that the outcome can be used to conduct SBO studies, where the cost of evaluating the QoI and derivatives is zero as we are working at the surrogate level. Ultimately, the choice of the method to use is to the user, and likely based on the computational resources available and in the difficulty to formulate the optimization problem.

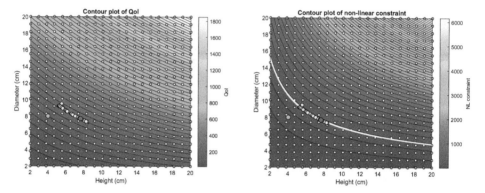

Figure 6. Left image: contour plot of the QoI (total surface). **Right image:** contour plot of the non-linear constraint (volume); in the image, the white line represents the range where the volume is 354 cm^3 < Volume < 356 cm^3. In both images, the green circle represents the starting point, the red circle represents optimal value, the yellow circles represent the path followed by the optimization algorithm (note that the gradient evaluations are not plotted), and the white circles represent the sampling points.

3.2. Static Mixer Optimization Case

In this case, we introduce the use of a qualitative metric to conduct the engineering design study. We also compare the outcome of a DO study and a DSE study. The geometry used in this case is shown in Figure 7, and it corresponds to a static mixer with two inlets and one output. The goal, in this case, is to obtain a given velocity distribution at the outlet by changing the angle of the inlet pipe 1 (refer to Figure 7). The velocity distribution field at the outlet was designed in such a way that the velocity normal to the outlet surface has a paraboloid distribution. Then, by using the SSIM index method (refer to Appendix A for an explanation), we compared the target image and the image of the current configuration (refer to Figure 8). The closer the SSIM index is to one, the more similar the images are; therefore, we aim at maximizing the QoI.

The simulations were conducted using OpenFOAM (version 7.0) [17,18]. To find the approximate solution of the governing equations, the SIMPLE pressure-velocity coupling method was used, together with the $k - epsilon$ turbulence model with wall functions, and a second-order accurate and stable discretization method for the convective, diffusive, and gradient terms. The Onshape document with all the dimensions is available at the following link (https://cad.onshape.com/documents/8f1312fafb3aac0f7bd3ed38/w/72a43b7cd8ca686e908ef122/e/33c606cd59a53e2b8532a94a).

The case setup is similar to the one presented in Section 3.1. The main difference is that we are introducing a new black-box application that requires additional steps so that it can be used inside the engineering design loop. The workflow specific to the data exchange between Dakota and the black-box solver (OpenFOAM in this case), is depicted in Figure 9 and discussed below. It is worth mentioning that the workflow is similar for different black-box applications, the only difference is in the formatting of the input and output files, and the data structure.

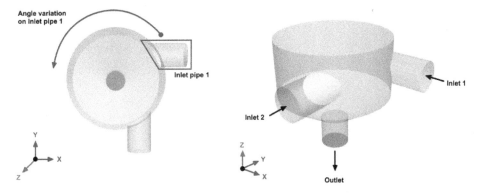

Figure 7. Static mixer geometry.

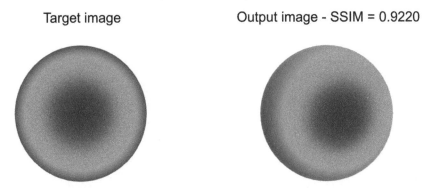

Figure 8. Velocity distribution normal to the outlet surface. **Left image:** reference velocity distribution or target image. **Right image:** image of the velocity distribution for a non-optimal case. To determine if the images are similar, we used the SSIM index method. The closer the SSIM index of the output image is to one, the more similar the images are.

First, the Dakota input file is setup to reflect the number, range, and name of the design variables (parametrical variables), the number of QoI, and the objective of the optimization study (minimize or maximize). In the same input file, the optimization method or design space method is chosen, along with the required options. Also, sequential or asynchronous function evaluations can be chosen according to the resources available. Then, as depicted in Figure 9, a **Template directory** is created to store the parametrical input files, i.e., subject to change as a result of the optimization process (e.g., files containing the definition of the geometry, boundary conditions for inlet velocity, physical properties, etc.). The automatic update of the parametrical files located in the **Template directory** is done automatically by using a Dakota supplied utility or user-defined scripts. These utilities skim all files located in the **Template directory** and automatically insert the values generated by Dakota during the design optimization or the design exploration study, into the predefined locations in the template files. In this workflow, a **Base case directory** is also created, where all the files needed to update the geometry and to run the OpenFOAM simulations are stored. The simulation control script file (or simulation driver), denoted by the **Control script** box in Figure 9, merges the automatically edited files in the **Template directory** with the **Base case directory**, creating in this way a working directory for a specific set of design parameters. At this point, the control script executes all the steps related to the simulation, i.e., geometry update, meshing, and launching the solver (in serial or parallel). Finally, all the data generated is automatically post-processed following the instructions defined in the control script. This includes quantitative and qualitative post-processing, as well as data formatting. It should be emphasized that the **Template directory** and **Base case directory** are created by the user. Also, the automatic update of the parametrical files is done after merging the directories **Template directory** and **Base case directory** into a separate working directory. For the interested reader, the working case setup can be found at this link (https://github.com/joelguerrero/cloud-based-cad-paper/tree/master/static_mixer).

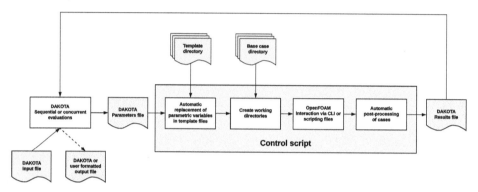

Figure 9. Workflow for data exchange between Dakota and OpenFOAM. The white rectangles denote process blocks, light-shaded blue document symbols denote unchanging sets of files, and light-shaded green document symbols indicate files that change with each set of design parameters generated by Dakota or after the end of the evaluation of the QoI. The light-shaded grey area denotes the domain of the control script that automatically prepares the case; this includes, CAD geometry, mesh generation, launching the solver, quantitative and qualitative post-processing, and automatic formatting of input and output files.

In Figure 10, we plot the outcome of the DO study using a gradient-based method (method of feasible directions or MFD [34,35] with numerical gradients computed using forward differences), and the DSE study using a uniform sampling for the inlet pipe angle (from 0 to 180 degrees). For the DO case, we used as starting point 0 degrees, and the case converged to the optimal value (pipe angle equal to 111.0549 degrees and SSIM index equal to 0.9660) in 31 function evaluations. In the DSE case, we explored the design space from 0 to 180 degrees, in steps of 5 degrees, so roughly speaking, we used the same number of function evaluations as for the DO case. From Figure 10, we can demonstrate that the DSE study, while not formerly converging to the optimal solution, gives more information about the design space than the DO method. From the DSE results, we can see that there is a plateau of the SSIM value for pipe angle values between 90 and 135 degrees. This information is not available when conducting DO studies, as the goal of these methods is to convergence to the optimal solution in an iterative fashion, and in doing so, some areas of the design space may remain unexplored. Using the data of the DSE study, we can also get a good estimate of the maximum value of the SSIM index, or we can use the data to construct a meta-model, and then use any DO method to find the optimal value. Both methods, DO and DSE, have their advantages and drawbacks and often is a good practice to use a combination of both, i.e., we first explore the design space in an inexpensive way, and then we use the information gathered from the DSE study to start a refined DO study.

In Figure 11, we show the velocity distribution at the outlet surface for five cases of the DSE study. In this figure, we also show the SSIM index value, the geometry layout, and the target image. As previously stated, the goal of this study was to obtain a given velocity distribution at the outlet (target image) by changing the angle of the inlet pipe. Then, by using the SSIM index method (Appendix A), we compare the target image and the image of the current configurations (as shown in Figure 11). The closer the SSIM index is to one, the more similar the images are. We highlight that we are using a qualitative metric instead of the traditional quantitative metrics used in engineering design studies. We designed beforehand the desired appearance of the field at the outlet, and then, by comparing the images in the design loop, we found the best match for our qualitative metric.

Again, we stress the fact that the loop is fully automatic and fault-tolerant, and it can be run concurrently and on the cloud. For the DSE case, we run eight simulations concurrently, each one using four cores. For the DO case, we were limited by the number of derivatives that can be computed at the same time. As this case only has one design variable, only one derivative can be computed. Therefore, the maximum number of concurrent simulations achievable in this DO case was two (one function evaluation and one gradient evaluation using forward differences), and each concurrent evaluation was conducted using eight cores.

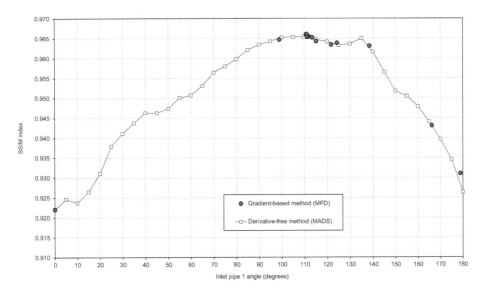

Figure 10. Comparison of the outcome of the DO and DSE studies. The QoI used was the SSIM index.

Let us now conduct a DSE study using three design variables, namely the diameter of the inlet pipe one, the diameter of the inlet pipe two, and the angle of the inlet pipe one. Again, all the parametrical variables were defined in the Onshape's document and modified using the Python API. This study was conducted using 170 experiments, generated using the space filling Latin hypercube sampling method (LHS) [2]. The simulations were run concurrently (eight simulations at the same time), and each simulation was run in parallel using four cores.

In Figure 12, we show another way to visualize high-dimensional data by using the parallel coordinates plot [43]. This kind of plot is extremely useful when visualizing and analyzing multivariate data, as it lets us identify how all variables are related. The highlighted line in Figure 12 represents the best solution (maximum SSIM index value), and shows the respective values of the design variables. In this DSE case, we can see that solutions that are better than the solution obtained using one design variable (SSIM = 0.9660), can be obtained by also changing the diameters of the inlet pipes. These solutions are shown in Figure 13. It worth mentioning that the parallel coordinates plots implemented are interactive; this allows us to isolate a range of values in real time. We can even change the order of the columns interactively and compare the slopes between variables. The scripts used for the parallel coordinates plots, as well as the data, are available at the following link (https://github.com/joelguerrero/cloud-based-cad-paper/tree/master/parallel_coordinates_dse_case). The interactive parallel coordinates plot can be accessed at the following link (http://joelguerrero.github.io/parallel_coordinates_dse_case/).

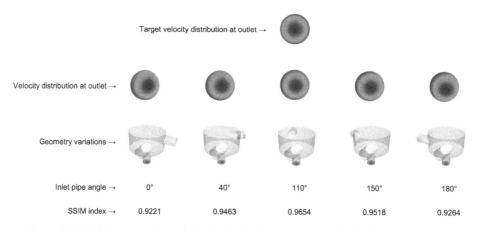

Figure 11. Qualitative comparison of the velocity distribution at the outlet. The SSIM method was used to compare the images. In the SSIM method, a value of 1 means that the images are identical. The target image is shown in the first row of the figure.

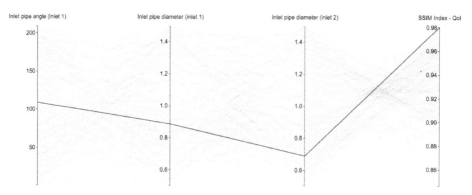

Figure 12. Parallel coordinates plot of the outcome of the DSE study using three design variables. The highlighted line represents the best solution.

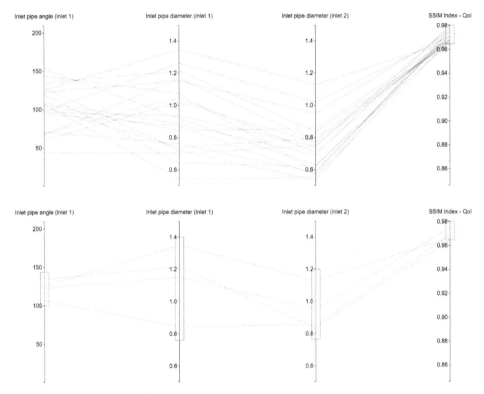

Figure 13. Parallel coordinates plot with filters. In the top image, the QoI has been filtered (0.9660 ≤ $SSIM$ ≤ 1). In the bottom figure, we apply additional filters to the design variables.

3.3. Two Ahmed Bodies in Platoon

In this case, we use the engineering design loop to conduct a parametric study. We compare the numerical results obtained with the current framework, against the experimental results obtained in references [44,45]; therefore, this is also a validation case. The simulations were conducted using OpenFOAM (version 7.0) [17,18]. To find the approximate solution of the governing equations, the SIMPLE pressure-velocity coupling method was used, together with the $k - \omega$ SST turbulence model with wall functions, and a second-order accurate and stable discretization method for the convective, diffusive, and gradient terms.

The study was conducted at different inter-vehicle spacing, an Ahmed body slant angle equal to 25 degrees, and an inlet velocity equal to 40 m/s. The QoI to measure is the normalized drag in platooning. In Figure 14, we depict a sketch of the computational domain and the definition of the inter-vehicle spacing S. From the parametrization used when creating the solid model, the two Ahmed bodies can be simulated in any formation with different slant angles, where everything can be controlled using configuration variables. The Onshape document with all the dimensions is available at the following link (https://cad.onshape.com/documents/b691f01f6fadba22433180ad/w/28165b21b45b4fee07e761b8/e/93c2ec3a1d01f9149d0557b1).

In Figure 15, we plot the outcome of this parametric study, where the normalized drag coefficient in platooning is computed as follows,

$$C_{D_{Platooning}} = \frac{C_{D1}}{C_{D2}} \qquad (9)$$

in this equation, C_{D1} is the drag coefficient of the Ahmed body in a platoon position (front, back, sideways, or any combination), and C_{D2} is the drag coefficient of the single Ahmed body. From the results presented in Figure 15, it can be observed a satisfactory agreement between the numerical and experimental values. It is worth mentioning that the simulations were run concurrently (four simulations at the same time), and each simulation was run in parallel using six cores.

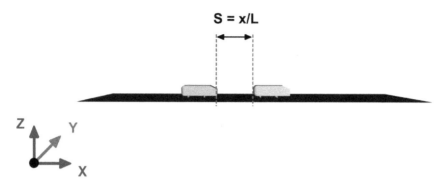

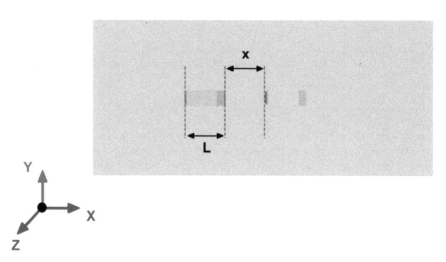

Figure 14. Spacing definition of the two Ahmed bodies, where x is the distance between the two bodies, L is the Ahmed body length, and S is the non-dimensional inter-vehicle spacing ($S = x/L$).

In this final application, we only conducted a parametrical study with one design variable. However, this study served to demonstrate the usability of the framework for complex validation cases. The reader should be aware that this case can be extended to more complex scenarios; for example, we could simulate one Ahmed body overtaking the other one.

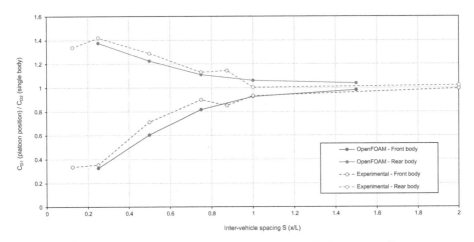

Figure 15. Normalized drag coefficient against inter-vehicle spacing S. The continuous lines represent the numerical results. The experimental results (dashed line) were taken from references [44,45].

4. Conclusions and Future Perspectives

In this manuscript, we presented an engineering design framework to perform design optimization and design space exploration studies. The engineering design loop implemented, allows for sequential and concurrent simulations (i.e., many simulations can be run at the same time), and each simulation can be run in parallel; this allows reduction of the output time of the design loop considerably. The optimization loop is fault-tolerant and software agnostic, and it can be interfaced with any application able to interact using input/output files via a command-line interface. The code coupling capabilities were provided by the library Dakota, and all the tools used in this work are open-source and freeware.

Two novel features were introduced in the workflow. First, the use of a cloud-based parametric CAD tool that gives engineers and designers complete control over the geometry during the design loop. This feature allows users to deploy the design loop in any platform as the installation is not required. It also lets the designers interact with the parametric CAD model using a programmatic API. Introducing the CAD tool into the design loop has been traditionally a problem because most of the CAD applications run in Windows OS. In contrast, the simulation software runs in Unix-like OS. Furthermore, in traditional CAD tools is not possible to interact with the parametric model using a programmatic environment; they take all the inputs via a graphical user interface that cannot be controlled in an automatic design loop. The use of the cloud-based CAD tool allowed us to circumvent these problems.

Secondly, the use of the SSIM index method to drive the design study. By using this metric, it is possible to compare images instead of integral quantities. We can now design beforehand how the field will look like in a given location of the domain, and the design loop will try to find the best match for that qualitative metric.

From the numerical experiments presented, it was demonstrated the flexibility and usability of the proposed workflow to tackle engineering design problems using different approaches. As for the optimization strategy concerns, we used gradient-based methods, derivative-free methods, surrogate-based optimization, and design space exploration techniques. All the methods delivered satisfactory results. The SSIM index method also proved to be very robust and easy to implement.

This tool, together with reduced-order models and surrogate models, has the potential to open the door to generative design in CFD. We look forward to working in this field, together with machine learning techniques and more advanced image recognition algorithms.

Author Contributions: Conceptualization, J.G.; Methodology, J.G., L.M. and S.B.N.; Software, J.G., L.M. and S.B.N.; Validation, J.G.; Formal Analysis, J.G.; Investigation, J.G., L.M. and S.B.N.; Resources, J.G.; Data Curation, J.G.; Writing—Original Draft Preparation, J.G. and L.M.; Writing—Review & Editing, J.G., L.M. and S.B.N.; Visualization, J.G.; Supervision, J.G. All authors have read and agreed to the published version of the manuscript.

Funding: This research received no external funding.

Acknowledgments: The use of the computing resources at CINECA high performance computing center was possible thanks to the ISCRA grant, project IsC45 DO4EnD2. We acknowledge the support provided by the AWS Cloud Credits for Research program. This work was conducted as part of the "Computational Optimization in Fluid Dynamics" course held by Jan Pralits and Joel Guerrero at the University of Genoa.

Conflicts of Interest: The authors declare no conflict of interest.

Appendix A

Hereafter, we briefly describe the Structural Similarity Index (SSIM) method used in Section 3.2 to measure the similarity between images. The SSIM is a method for predicting the perceived quality of digital television and cinematic pictures, as well as other kinds of digital images and videos.

Referring to a grey-scale image, a similarity index can be computed considering it as a bi-dimensional function of intensity [46]. The simplest and most commonly used similarity index is the mean squared error (MSE), which is obtained averaging the squared intensity difference between two pictures on each pixel [47]. However, the MSE, like many other mathematically defined indexes, is not able to take into account subjective quality measures (i.e., human perception-based criteria, such as image structure comparison) [48]. For this reason, it can be misleading when it is necessary to find the image that is more similar to a reference one.

To avoid the problems related to the MSE, the SSIM index can be used. Based on how it is defined, the SSIM takes into account the structured information and the neighborhood dependencies that are usually present in natural images. The SSIM has been used with success in different research fields; for example, in reference [49], the authors used it to detect disturbances or blurring effects in a set of pictures. The authors also reported that it was not possible to do the same with the MSE. In reference [50], the SSIM index of flame images was used as a measure of the burning state in a sintering process. By using a small number of samples, the authors were able to recognize the burning state with satisfactory accuracy thanks to the SSIM index. In reference [51], a hand gesture recognition study based on both MSE and SSIM was presented, and it was concluded that both techniques could be used for gesture recognition. In addition, it was also found that the SSIM was superior to the MSE, as it was insensitive to small imperfections in the reconstructed image caused by thresholding.

Considering two different image discrete signals, let us say x and y, the similarity evaluation is based on three characteristics: luminance, contrast, and structure [47]. The luminance μ_x of each signal is computed as the mean intensity, as follows,

$$\mu_x = \frac{1}{N} \sum_{i+1}^{N} x_i \tag{A1}$$

where N is the number of pixels.

The luminance comparison between x and y is then performed defining the function $l(x,y)$,

$$l(x,y) = \frac{2\mu_x\mu_y + C_1}{\mu_x^2 + \mu_y^2 + C_1} \tag{A2}$$

where C_1 is a constant used to avoid instabilities when the denominator is close to zero.

The contrast σ_x is estimated as the standard deviation of the image signal, and is computed as follows,

$$\sigma_x = \sqrt{\frac{1}{N-1} \sum_{i+1}^{N} (x_i - \mu_x)^2} \tag{A3}$$

The contrast comparison function $c(x,y)$ is similar to Equation (A2), and it also includes a constant to avoid instabilities (C_2).

$$c(x,y) = \frac{2\sigma_x \sigma_y + C_2}{\sigma_x^2 + \sigma_y^2 + C_2} \tag{A4}$$

The structure comparison can be performed by defining the function $s(x,y)$,

$$s(x,y) = \frac{\sigma_{xy} + C_3}{\sigma_x \sigma_y + C_3} \tag{A5}$$

where σ_{xy} is specified as follows,

$$\sigma_{xy} = \frac{1}{N-1} \sum_{i+1}^{N} (x_i - \mu_x)(y_i - \mu_y) \tag{A6}$$

Finally, by combining Equations (A2), (A4) and (A5), it is possible to obtain the SSIM index between x and y, as follows,

$$SSIM(x,y) = [l(x,y)]^\alpha \cdot [c(x,y)]^\beta \cdot [s(x,y)]^\gamma \tag{A7}$$

where α, β, and γ are positive parameters used as weights factors to set the importance of $l(x,y)$, $c(x,y)$ and $s(x,y)$ when computing the SSIM index. A simplified expression of Equation (A7) can be obtained by setting $l(x,y)$, $c(x,y)$, $s(x,y)$, and C_3 to the following values [47],

$$\alpha = 1 \qquad \beta = 1 \qquad \gamma = 1 \qquad C_3 = \frac{C_2}{2} \tag{A8}$$

thus obtaining the following expression for SSIM (which is the form of the Equation (A7) used in this work),

$$SSIM(x,y) = \frac{(2\mu_x \mu_y + C_1)(2\sigma_{xy} + C_2)}{\left(\mu_x^2 + \mu_y^2 + C_1\right)\left(\sigma_x^2 + \sigma_y^2 + C_2\right)} \tag{A9}$$

To analyze the images, we use the Python library scikit-image [52], which is a collection of algorithms for image processing. The images to compare are saved as color images in digital format (e.g., Portable Network Graphics or PNG format). However, this procedure was designed for grey-scale images, as stated at the beginning of this section. Thus, it is necessary to separate the three different color channels (red, green, and blue), as shown in Figure A1. This is done by using the Python function **imread** to import the digital image (in PNG format) as a **uint8** three-dimensional array. At this point, each channel is a monochrome picture so that it can be treated as a grey-scale picture, and its SSIM index can be computed by using Equation (A9). The SSIM of the original digital image can be finally obtained as the average of the SSIMs of the three color channels. The computation of the SSIM of the separate channels and their averaging is performed using the **compare_ssim** function implemented in the Python library scikit-image. The SSIM index value is a number between 0 and 1, where 1 means a perfect matching between the images. That is, the closer the value is to 1, the more similar the images are. A sample python script can be found at the following link (https://github.com/joelguerrero/cloud-based-cad-paper/tree/master/SSIM).

Figure A1. Separation of red, green and blue channels of a color picture. Image courtesy of Diego Rattazzi (diego.rattazzi@edu.unige.it).

References

1. Mattson, C.A. Design Exploration. Available online: https://design.byu.edu/blog/design-exploration-presentation-given-stanford-university-15-jan-2014 (accessed on 22 February 2020).
2. Forrester, A.; Sobester, A.; Keane, A. *Engineering Design via Surrogate Modeling. A Practical Guide*; Wiley: Hoboken, NJ, USA, 2008.
3. Guerrero, J.; Cominetti, A.; Pralits, J.; Villa, D. Surrogate-Based Optimization Using an Open-Source Framework: The Bulbous Bow Shape Optimization Case. *Math. Comput. Appl.* **2018**, *23*, 60. [CrossRef]
4. Romero, V.J.; Swiler, L.P.; Giunta, A.A. Construction of response surfaces based on progressive lattice-sampling experimental designs. *Struct. Saf.* **2004**, *26*, 201–219. [CrossRef]
5. Kochenderfer, M.; Wheeler, T. *Algorithms for Optimization*; MIT Press: Cambridge, MA, USA, 2019.
6. Gen, M.; Cheng, R. *Genetic Algorithms and Engineering Optimization*; Wiley-Interscience: Hoboken, NJ, USA, 2000.
7. Vanderplaats, G. *Multidiscipline Design Optimization*; Vanderplaats Research & Development, Inc.: Colorado Springs, CO, USA, 2007.
8. Papalambros, P.; Wilde, D. *Principles of Optimal Design. Modeling and Computation*; Cambridge University Press: Cambridge, UK, 2017.
9. Nocedal, J.; Wright, S. *Numerical Optimization*; Springer: Berlin/Heidelberg, Germany, 2006.
10. Chong, E.; Zak, S. *An Introduction to Optimization*; Wiley: Hoboken, NJ, USA, 2013.
11. Sobieszczanski-Sobieski, J.; Morris, A.; van Tooren, M.; Rocca, G.L.; Yao, W. *Multidisciplinary Design Optimization Supported by Knowledge Based Engineering*; Wiley: Hoboken, NJ, USA, 2015.
12. Chauhan, S.; Hwang, J.; Martins, J. An automated selection algorithm for nonlinear solvers in MDO. *Struct. Multidiscip. Optim.* **2018**, *58*, 349–377. [CrossRef]
13. Martins, J. The adjoint method in multidisciplinary design optimization—Special session in honor of Antony Jameson's 85th birthday. In Proceedings of the AIAA SciTech Forum, Nashville, TN, USA, 6 January 2020.
14. Vassberg, J.C.; Jameson, A. *Introduction to Optimization and Multidisciplinary Design Part I: Theoretical Background for Aerodynamic Shape Optimization*; Lecture Series March 2016; Von Karman Institute: Brussels, Belgium, 2016.
15. Keane, A.J.; Nair, P.B. *Computational Approaches for Aerospace Design: The Pursuit of Excellence*; John Wiley & Sons: Hoboken, NJ, USA, 2005.
16. Martins, J.R.R.A.; Lambe, A.B. Multidisciplinary Design Optimization: A Survey of Architectures. *AIAA J.* **2013**, *51*, 2049–2075. [CrossRef]
17. The OpenFOAM Foundation. Available online: http://www.openfoam.org (accessed on 22 February 2020).
18. Weller, H.G.; Tabor, G.; Jasak, H.; Fureby, C. A tensorial approach to computational continuum mechanics using object-oriented techniques. *Comput. Phys.* **1998**, *12*, 620–631. [CrossRef]
19. Dakota Web Page. 2018. Available online: https://dakota.sandia.gov/ (accessed on 22 February 2020).

20. Adams, B.M.; Eldred, M.S.; Geraci, G.; Hooper, R.W.; Jakeman, J.D.; Maupin, K.A.; Monschke, J.A.; Rushdi, A.A.; Stephens, J.A.; Swiler, L.P.; et al. *Dakota, a Multilevel Parallel Object-Oriented Framework for Design Optimization, Parameter Estimation, Uncertainty Quantification, and Sensitivity Analysis: Version 6.10 User Manual*; Sandia National Laboratories: Albuquerque, NM, USA, 2019.
21. The Visualization Toolkit (VTK). Available online: http://www.vtk.org (accessed on 22 February 2020).
22. Onshape Product Development Platform. Available online: http://www.onshape.com (accessed on 22 February 2020).
23. Slotnick, J.; Khodadoust, A.; Alonso, J.; Darmofal, D.; Gropp, W.; Lurie, E.; Mavriplis, D. *CFD Vision 2030 Study: A Path To Revolutionary Computational Aerosciences*; Technical Report; NASA: Hampton, VA, USA, 2014.
24. Daymo, E.; Tonkovich, A.L.; Hettel, M.; Guerrero, J. Accelerating reactor development with accessible simulation and automated optimization tools. *Chem. Eng. Process.-Process Intensif.* **2019**, *142*, 107582. [CrossRef]
25. Xia, C.C.; Gou, Y.J.; Li, S.H.; Chen, W.F.; Shao, C. An Automatic Aerodynamic Shape Optimisation Framework Based on DAKOTA. *IOP Conf. Ser. Mater. Sci. Eng.* **2018**, *408*, 012021. [CrossRef]
26. Byrne, J.; Cardiff, P.; Brabazon, A.; O'Neill, M. Evolving parametric aircraft models for design exploration. *J. Neurocomput.* **2014**, *142*, 39–47. [CrossRef]
27. Ohm, A.; Tetursson, H. *Automated CFD Optimization of a Small Hydro Turbine for Water Distribution Networks*; Technical Report; Chalmers University of Technology: Göteborg, Sweden, 2017.
28. Sousa, J.; Gorlé, C. Computational urban flow predictions with Bayesian inference: Validation with field data. *Build. Environ.* **2019**, *154*, 13–22. [CrossRef]
29. Kiani, H.; Karimi, F.; Labbafi, M.; Fathi, M. A novel inverse numerical modeling method for the estimation of water and salt mass transfer coefficients during ultrasonic assisted-osmotic dehydration of cucumber cubes. *Ultrason. Sonochem.* **2018**, *44*, 171–176. [CrossRef]
30. Habla, F.; Fernandes, C.; Maier, M.; Densky, L.; Ferras, L.; Rajkumar, A.; Carneiro, O.; Hinrichsen, O.; Nobrega, J.M. Development and validation of a model for the temperature distribution in the extrusion calibration stage. *Appl. Therm. Eng.* **2016**, *100*, 538–552. [CrossRef]
31. Khamlaj, T.A.; Rumpfkeil, M.P. Analysis and optimization of ducted wind turbines. *Energy* **2018**, *162*, 1234–1252. [CrossRef]
32. Montoya, M.C.; Nieto, F.; Hernandez, S.; Kusano, I.; Alvarez, A.; Jurado, J. CFD-based aeroelastic characterization of streamlined bridge deck cross-sections subject to shape modifications using surrogate models. *J. Wind Eng. Ind. Aerodyn.* **2018**, *177*, 405–428. [CrossRef]
33. Kelm, S.; Müller, H.; Hundhausen, A.; Druska, C.; Kuhr, A.; Allelein, H.J. Development of a multi-dimensional wall-function approach for wall condensation. *Nucl. Eng. Des.* **2019**, *353*, 110239. [CrossRef]
34. Zoutendijk, G. *Methods of Feasible Directions: A Study in Linear and Non-Linear Programming*; Elsevier: Amsterdam, The Netherlands, 1960.
35. Vanderplaats, G.N. An efficient feasible directions algorithm for design synthesis. *AIAA J.* **1984**, *22*, 1633–1640. [CrossRef]
36. Le Digabel, S. Algorithm 909: NOMAD: Nonlinear Optimization with the MADS Algorithm. *ACM Trans. Math. Softw.* **2011**, *37*. [CrossRef]
37. Guerrero, J. Opportunities and challenges in CFD optimization: Open Source technology and the Cloud. In Proceedings of the Sixth Symposium on OpenFOAM® in Wind Energy (SOWE), Göteborg, Sweden, 13–14 June 2018.
38. Oliver, M.; Webster, R. Kriging: A Method of Interpolation for Geographical Information Systems. *Int. J. Geogr. Inf. Syst.* **1990**, *4*, 313–332. [CrossRef]
39. Adams, B.M.; Eldred, M.S.; Geraci, G.; Hooper, R.W.; Jakeman, J.D.; Maupin, K.A.; Monschke, J.A.; Rushdi, A.A.; Stephens, J.A.; Swiler, L.P.; et al. *Dakota, a Multilevel Parallel Object-Oriented Framework for Design Optimization, Parameter Estimation, Uncertainty Quantification, and Sensitivity Analysis: Version 6.10 Theory Manual*; Sandia National Laboratories: Albuquerque, NM, USA, 2014.
40. Dalbey, K.R.; Giunta, A.A.; Richards, M.D.; Cyr, E.C.; Swiler, L.P.; Brown, S.L.; Eldred, M.S.; Adams, B.M. *Surfpack User's Manual Version 1.1*; Sandia National Laboratories: Albuquerque, NM, USA, 2013.
41. Dalbey, K.R. *Efficient and Robust Gradient Enhanced Kriging Emulators*; Technical Report; Sandia National Laboratories: Albuquerque, NM, USA, 2013.

42. Giunta, A.A.; Watson, L. A comparison of approximation modeling techniques: Polynomial versus interpolating models. In *7th AIAA/USAF/NASA/ISSMO Symposium on Multidisciplinary Analysis and Optimization*; NASA Langley Technical Report Server: Hampton, VA, USA, 1998.
43. Inselberg, A. *Parallel Coordinates, Visual Multidimensional Geometry and Its Applications*; Springer: Berlin/Heidelberg, Germany, 2009.
44. Pagliarella, R.M.; Watkins, S.; Tempia, A. Aerodynamic Performance of Vehicles in Platoons: The Influence of Backlight Angles. *SAE World Congr. Exhib. SAE Int.* **2007**. [CrossRef]
45. Pagliarella, R.M. *On the Aerodynamic Performance of Automotive Vehicle Platoons Featuring Pre and Post-Critical Leading Forms*; Technical Report; RMIT University: Melbourne, Austrialia, 2009.
46. Sampat, M.P.; Wang, Z.; Gupta, S.; Bovik, A.C.; Markey, M.K. Complex wavelet structural similarity: A new image similarity index. *IEEE Trans. Image Process.* **2009**, *18*, 2385–2401. [CrossRef]
47. Wang, Z.; Bovik, A.C.; Sheikh, H.R.; Simoncelli, E.P. Image Quality Assessment: From Error Visibility to Structural Similarity. *IEEE Trans. Image Process.* **2004**, *13*, 600–612. [CrossRef]
48. Eskicioglu, A.M.; Fisher, P.S. Image Quality Measures and Their Performance. *IEEE Trans. Commun.* **1995**, *43*, 2959–2965. [CrossRef]
49. Ndajah, P.; Kikuchi, H.; Yukawa, M.; Watanabe, H.; Muramatsu, S. SSIM image quality metric for denoised images. In Proceedings of the International Conference on Visualization, Imaging and Simulation, Faro, Portugal, 3–5 November 2010; pp. 53–57.
50. Lin, Y.; Chai, L.; Zhang, J.; Zhou, X. On-line burning state recognition for sintering process using SSIM index of flame images. In Proceedings of the 11th World Congress on Intelligent Control and Automation, Shenyang, China, 29 June–4 July 2014; pp. 2352–2357. [CrossRef]
51. Priyal, S.P.; Bora, P.K. A study on static hand gesture recognition using moments. In Proceedings of the International Conference on Signal Processing and Communications (SPCOM), Bangalore, India, 18–21 July 2010; pp. 1–5. [CrossRef]
52. Van der Walt, S.; Schönberger, J.L.; Nunez-Iglesias, J.; Boulogne, F.; Warner, J.D.; Yager, N.; Gouillart, E.; Yu, T. Scikit-image: Image processing in Python. *PeerJ* **2014**, *2*, e453. [CrossRef]

© 2020 by the authors. Licensee MDPI, Basel, Switzerland. This article is an open access article distributed under the terms and conditions of the Creative Commons Attribution (CC BY) license (http://creativecommons.org/licenses/by/4.0/).

MDPI
St. Alban-Anlage 66
4052 Basel
Switzerland
Tel. +41 61 683 77 34
Fax +41 61 302 89 18
www.mdpi.com

Fluids Editorial Office
E-mail: fluids@mdpi.com
www.mdpi.com/journal/fluids

CPSIA information can be obtained
at www.ICGtesting.com
Printed in the USA
LVHW070845300720
661935LV00036B/1541